Universitext

Springer
New York
Berlin
Heidelberg
Barcelona
Budapest
Hong Kong
London
Milan
Paris
Santa Clara
Singapore
Tokyo

Vivek S. Borkar

Probability Theory

An Advanced Course

Vivek S. Borkar
Department of Electrical Engineering
Indian Institute of Science
Bangalore 560012
India

Mathematics Subject Classification (1991): 60-02

Library of Congress Cataloging-in-Publication Data
Borkar, Vivek S.
Probability theory: an advanced course / Vivek S. Borkar.
p. cm. - (Universitext)
Includes bibliographical references (p. -) and index.
ISBN 0-387-94558-X (softcover : a1k. paper)
1. Probabilities. I. Title.
QA273.B823 1995
519.2-dc20 95-34388

ISBN 0-387-94558-X Springer-Verlag New York Berlin Heidelberg

Borkar: Probability Theory: An Advanced Course

First Indian Reprint 2012

ISBN 978-81-322-0788-7

This edition is published by Springer (India) Private Limited, (a part of Springer Science+Business Media), Registered Office: 7th Floor, Vijaya Building, 17 Barakhamba Road, New Delhi - 110 001, India

Printed in India by Rakmo Press Pvt. Ltd., New Delhi.

Preface

This book presents a selection of topics from probability theory. Essentially, the topics chosen are those that are likely to be the most useful to someone planning to pursue research in the modern theory of stochastic processes. The prospective reader is assumed to have good mathematical maturity. In particular, he should have prior exposure to basic probability theory at the level of, say, K.L. Chung's 'Elementary probability theory with stochastic processes' (Springer-Verlag, 1974) and real and functional analysis at the level of Royden's 'Real analysis' (Macmillan, 1968).

The first chapter is a rapid overview of the basics. Each subsequent chapter deals with a separate topic in detail. There is clearly some selection involved and therefore many omissions, but that cannot be helped in a book of this size. The style is deliberately terse to enforce active learning. Thus several tidbits of deduction are left to the reader as labelled exercises in the main text of each chapter. In addition, there are supplementary exercises at the end.

In the preface to his classic text on probability ('Probability', Addison-Wesley, 1968), Leo Breiman speaks of the right and left hands of probability. To quote him: "On the right is the rigorous foundational work using the tools of measure theory. The left hand 'thinks probabilistically', reduces problems to gambling situations, coin-tossing, motions of a physical particle." This is a right-handed book, though a brief "prologue" has been inserted to give an inkling about the missing left hand. An ambidextrous book would be huge and also very difficult to write (at least for the present author). The reader is warned of this shortcoming and is strongly advised to acquire the "left hand" on his own through supplementary reading. Not having it is a major handicap. Also, it's no fun.

Needless to say, I did not invent this subject. Thus barring some novelty of organization and occasional variations on usual proofs, the material is standard. My rendition of it has been heavily influenced by two factors. The first is a beautiful stream of probability courses I took at Berkeley, taught variously by Professors David Blackwell, Michael Klass, Aram Thomasian, J.W. Pitman and David Aldous. (In particular, Chapter VI owes a lot to a course I took from Prof. Aldous.) The second is the several texts I "grew up with" during the formative years as a graduate student. These include, in addition to Breiman's book mentioned above, 'A first course in probability theory' by K.L. Chung (Academic, 1974), 'Probability theory — independence, interchangeability and martingales' by Y.S. Chow and H. Teicher (Springer-Verlag, 1978), 'Discrete parameter martingales', by J. Neveu (North Holland, 1975), 'Convergence of probability measures' by P. Billingsley (Wiley, 1968) and 'Probabilities and potential' by C. Dellacherie and P.A. Meyer (North Holland, 1978). I have also used 'Probability and measure' by P. Billingsley (Wiley, 1979) and 'Introduction to probability and measure' by K.R. Parthasarathy (Macmillan (India), 1977).

Several people have contributed to bringing this book about. A major credit goes to Dr. V.V. Phansalkar who contributed a lot to the "clean up" operations at various stages. Dr. P.G. Babu, Dr. P.S. Sastry and Mr. G. Santharam also helped in several ways. The financial burden was borne by a generous grant from the Curriculum Development Cell of the Centre for Continuing Education, Indian Institute of Science, for which I am extremely grateful.

Writing this book has been my pet project for quite a while. I hope that in its final execution I have done justice to myself and to its potential readers.

This book is dedicated to the memory of my father-in-law, the late Shri Manohar N. Budkule.

Prologue

"The equanimity of your average tosser of coins depends upon the law, or rather a tendency, or let us say a probability, or at any rate a mathematically calculable chance, which ensures that he will not upset himself by losing too much nor upset his opponent by winning too often. This made for a kind of harmony and a kind of confidence. It related the fortuitous and the ordained into a reassuring union which we recognized as nature."
— Guildenstern in Tom Stoppard's *Rosencrantz and Guildenstern Are Dead* (Faber and Faber Ltd, London, 1967).

What is probability? It is not easy to answer this question. At the level of gut feeling, one can hardly better the above quote from Stoppard. Trying to go any deeper into the definition of probability will quickly get us into the realms of logic and philosophy. (See, e.g., [24].) We shall evade this issue altogether by taking a phenomenological viewpoint. Take, for instance, the simplest and the most quoted probabilistic phenomenon — the tossing of a coin. Consider the pair of "cause" (tossing a coin) and "effect" (coin drops on the floor) that goes with it. This is as clear an instance of a cause-effect relationship as one would wish to have. But refine the "effect" a little, say, by replacing it by "coin drops with head up" and we are already in trouble. This is only one of the possible effects since the coin could have equally well settled down with tail up. Which of these possibilities will occur? We cannot say a priori. But our intuition suggests that they are equally likely. Of course, another person with a different kind of intuition (or a better knowledge of the coin's composition) could think otherwise. But even after granting that it is a subjective judgement, how do we justify it? One way to do so would be to say that in absence of any extra information, there is no reason to prefer one outcome over the other ("the principle of insufficient reason"). Alternatively, we may take a purely empirical viewpoint, viz., toss the coin several times and verify that both head and tail come up in approximately half the instances. The latter already suggests a way to quantify our judgement — we say that head and tail have equal probability of one half each.

One may say all this to a *random* man on the street and in all *probability*, he won't find anything spurious with this argument. But wait — there is a *chance* that he is a sceptic and may say that the "probability" in the coin-tossing experiment is only apparent. If one finds out exactly all the parameters affecting the coin's motion (its initial coordinates, orientation, direction and magnitude of the thrust, details of the surface it falls upon, etc.) one can exactly predict the outcome. This is certainly a valid objection, but this kind of reasoning applied to probabilistic phenomena has its limitations. There may be limitations to measurement of these parameters, either practical or fundamental (e.g., quantum mechanical or "computational complexity" based). Even when they are not there, it may be easier (quicker, cheaper, etc.) to model and analyze the phenomenon as probabilistic. Without getting into the details of these issues, we shall accept probability as a tried and tested paradigm and tool for modelling and analysis of certain phenomena, viz., those in which a cause can lead to one of many outcomes (for whatever reasons) to each of which a quantitative measure of comparative likelihood can be assigned (in whatever manner). Essentially, the reader hereby is being asked to accept the "gut feeling" we started with in lieu of a definition. Having done so, let us try to mathematicize the concept.

Going back to coin-tossing, consider the set of all possible values of the parameters that determine the coin's motion. We shall call this the "sample space", denoted by Ω. Our act of tossing the coin at a particular time in a particular surrounding in a particular manner picks ("samples") a point from this space. This may either fall into the set $A = \{\text{points of } \Omega \text{ that lead to head}\}$ or $A^c = \{\text{points of } \Omega \text{ that lead to tail}\}$. (Note that I am already discounting the "improbable" events such as the coin standing on its edge.) This partitions Ω into two subsets, to each of which our "gut feeling" assigns a probability $\frac{1}{2}$. Next, consider a slightly more complicated situation, say, the rolling of a die. The new sample space Ω now partitions into six subsets $A_i, 1 \leq i \leq 6$, with $A_i = \{\text{sample points that lead to number } i\}$ for each i. Again, our intuition assigns a probability of $\frac{1}{6}$ to each A_i. Moreover, it makes sense to talk of the probability of getting an even number = the probability of $A_2 \bigcup A_4 \bigcup A_6 = (\frac{1}{6}) + (\frac{1}{6}) + (\frac{1}{6}) = \frac{1}{2}$, and so on.

More generally, one has a set Ω called the sample space and a collection $\mathcal{F}$ of its subsets called events. Elements of Ω are called sample points. To each event A is assigned a number between zero and one, called its probability and denoted by $P(A)$. Now we expect ("gut feeling" again) that $\mathcal{F}$ and $P(.)$ as a map from $\mathcal{F}$ to [0,1] should satisfy certain requirements. For example, "nothing happens" (empty set ϕ) and "something happens" (Ω) should be events. If A is an event, A^c(= "A does not occur") should also be one. If $A_1, A_2, A_3, \cdots$ are events, "at least one of the A_i's occurs" ($\bigcup A_i$) and "all of $\{A_i\}$ occur" ($\bigcap A_i$) should also be events. In short, $\mathcal{F}$ is a σ-field. As for $P(.), P(\Omega) = 1$ by convention. (This can also be rationalized via the "relative frequency of occurrence" interpretation.) Of course, $P(\phi)$ has to be zero and $P(A^c) = 1 - P(A)$. Moreover, if $A_1, A_2, A_3, \cdots$, are disjoint, it makes sense to demand that $P(\bigcup A_i) = \Sigma P(A_i)$. In other words, P is a countably additive nonnegative measure on the measurable space $(\Omega, \mathcal{F})$ with total mass 1. We call such a measure a probability measure.

Now an experiment such as tossing a coin or rolling a die picks a point ω from the sample space and maps it into an element of another space E. (E = {head, tail} and {1,2,3,4,5,6} resp. in the two examples.) Thus it is a map $X : \Omega \to E$. Since our idea is to have sets of the type $\{\omega \mid X(\omega) \in B\}$ to be events (i.e., elements of $\mathcal{F}$) for a suitable collection of subsets $B \subset E$, considerations analogous to the above for $\mathcal{F}$ suggest that we equip E with a σ-field ξ and require the map $X : (\Omega, \mathcal{F}) \to (E, \xi)$ to be measurable. Such a map will be called an (E-valued) random variable. Our mathematical formalism for probability theory is now ready.

Having glibly said all this, let me hasten to add that none of it is obvious. The mathematical formulation of probability theory has a long history and was, in fact, among the major open issues in mathematics early this century. Its eventual settlement via the measure theoretic framework as sketched above is due to Kolmogorov and followed a lot of early work by several others such as Markov and Borel. There were other contenders too, such as the "relative frequency" approach of Von Mises. By now, Kolmogorov's formulation is the most widely accepted one in the mathematics community and we shall stick to it. The foundational issues, however, are by no means dead, one of the most contested issues being the hypothesis of countable additivity. See [43] for a recent debate on these matters. Readers interested in foundational issues should look up [20, 41]. For historical details, see [10, 14, 33].

Note that in reality one observes only a single "realization" of a random variable X, i.e., $X(\omega)$ for a particular $\omega \in \Omega$. Thus the probability space $(\Omega, \mathcal{F}, P)$ in the background is a hypothetical entity and its choice is by no means unique. There is also the problem of choosing P. Some methods thereof are listed below.

(i) *Principle of insufficient reason* — In absence of any reason to favour one outcome over the other, we may deem them equally likely. This can be made a building block for deriving more complicated P. For example, the observable outcome may be a known function of more basic variables to which this principle applies. This function need not be one-one and thus the possible values of the observed outcome need not be equally likely. This is the basis of the Darwin–Fowler approach to statistical mechanics [30, Chapter 5].

(ii) *Subjective probability* — P may simply be a quantification of one's subjective beliefs regarding the relative likelihoods of various events and thus liable to change from person to person. These considerations are important in economic applications [41].

(iii) *Physical reasoning* — Some knowledge of the underlying physical phenomena coupled with simplifying assumptions and probabilistic reasoning can lead to a natural choice of P in some cases. For example, noise in electric circuits is often the cumulative effect of many small and essentially independent phenomena. The central limit theorem (Chapter IV) then suggests that it may be taken to be Gaussian.

(iv) *Worst case analysis* — When a random effect is unwanted (e.g. noise), the "worst case analysis" approach suggests that we take the P that is the worst in an appropriate sense. The "maximum entropy" method in statistics is based on such considerations [26].

(v) *Mathematical simplicity* — One often hypothesizes a P for its mathematical simplicity. The ensuing analysis should in principle be justified by suitable "robustness" results which show that the deductions remain essentially valid even when the P is perturbed.

(vi) *Measurements* — In repeatable random phenomena, one may estimate P from measurements. This is what "Statistics" is all about.

Probability theory has had a symbiotic relationship with several other disciplines and no essay on probability would be complete without at least a mention of these. They are:

(1) Engineering ("noise" in communication engineering, random vibrations in structural engineering),

(2) Operations research and computer science (queuing models, stochastic search algorithms),

(3) Physics (statistical and quantum mechanics, astrophysics),

(4) Biology (genetics, population dynamics),

(5) Economics (econometrics, information economics),

(6) Other branches of mathematics (ergodic theory, partial differential equations),

and so on.

With this we conclude our preamble and move on to the mathematical theory of probability, which begins with $(\Omega, \mathcal{F}, P)$.

Contents

1

Introduction

1.1 Random Variables

Let $(\Omega, \mathcal{F}, P)$ be a probability space. To recapitulate: Ω is a set called the "sample space". Its elements are called sample points. $\mathcal{F}$ is a σ-field of subsets of Ω containing Ω itself. Elements of $\mathcal{F}$ are called events. P is a probability measure (i.e., a countably additive nonnegative measure with total mass 1) on the measurable space $(\Omega, \mathcal{F})$. If an event A is of the type $A = \{\omega \in \Omega \mid R(\omega)\}$ for some property $R(.)$, we may write $P(R)$ for $P(A)$. An event is called a sure event if $P(A) = 1$ and a null event if $P(A) = 0$. Alternatively, $R(.)$ is said to hold almost surely (a.s. for short) if $P(R) = 1$. Many statements in probability theory are made with the qualification "almost surely", though this may not always be stated explicitly.

Let (E, ξ) be another measurable space and $X : (\Omega, \mathcal{F}, P) \rightarrow (E, \xi)$ a random variable (i.e., a measurable map). The image μ of P under X is a probability measure on (E, ξ), called the law of X and denoted by $\mathcal{L}(X)$. The events $\{\omega \mid X(\omega) \in A\}$ for $A \in \xi$ form a sub-σ-field of $\mathcal{F}$ called the σ-field generated by X and denoted by $\sigma(X)$. More generally, given a family $X_\alpha, \alpha \in I$, of random variables on $(\Omega, \mathcal{F}, P)$ taking values in measurable spaces $(E_\alpha, \xi_\alpha), \alpha \in I$, respectively, the σ-field generated by $X_\alpha, \alpha \in I$, denoted by $\sigma(X_\alpha, \alpha \in I)$, is the smallest sub-$\sigma$-field with respect to which they are all measurable. There may be situations where it is preferable to view $\{X_\alpha, \alpha \in I\}$ as a single random variable taking values in the product space $\prod E_\alpha$ endowed with the product σ-field $\prod \xi_\alpha$. If so, this definition reduces to the preceding one.

Two (or more) random variables are said to agree in law if their laws coincide. They could be defined on different probability spaces. For example, in the above set-up, the random variable $Y : (E, \xi, \mu) \to (E, \xi)$ defined by $Y(\omega) = \omega$ agrees in law with X. This special choice of probability space is said to be "canonical" and Y is said to be "canonically realized".

In probability theory, a certain class of (E, ξ) merits special attention. These are the Polish spaces (i.e., separable topological spaces that permit a complete metrization) endowed with their Borel σ-fields. Examples are: separable Banach spaces, $C([0, \infty))$ and $D([0, \infty))$ (resp. the space of continuous functions from $[0, \infty)$ to $\boldsymbol{R}$ and the space of right-continuous functions from $[0, \infty)$ to $\boldsymbol{R}$ for which left limits exist, with appropriate topologies), the space of probability measures on a Polish space with the Prohorov topology introduced in the next chapter. The following theorem states an important property of these spaces. Say that two measurable spaces are measurably isomorphic if there exists a bijection between them which is measurable in both directions.

Theorem 1.1.1 *A Polish space S is homeomorphic to a G_δ subset of $[0, 1]^\infty$ and measurably isomorphic to a measurable subset of $\boldsymbol{R}$.*

Proof Let d be a complete metric on S taking values in $[0, 1)$. (If it does not, replace it by $d(.,.)/(1 + d(.,.))$). Let $\{s_n\} \subset S$ be countable dense. Define $h : S \to [0, 1]^\infty$ by $h(x) = [d(x, s_1), d(x, s_2), \cdots]$. Then h is continuous. Also, $x_n \not\to x$ in S implies $d(x_n, x) > \epsilon$ infinitely often (i.o.) for some $\epsilon > 0$. Thus for an s_k with $d(x, s_k) < \epsilon/2, d(x_n, s_k) > \epsilon/2$ i.o., implying $h(x_n) \not\to h(x)$. Thus h is a homeomorphism between S and $h(S)$. Let $\bar{d}$ be a metric on $[0, 1)^\infty$. For each $h(x), x \in S$, take $B \subset [0, \infty)^\infty$ open such that $h(x) \in B$ and the diameters of B and $h^{-1}(B)$ are less than $1/m$ for a prescribed $m \geq 1$. Let G_m be the union of all such B. Then $h(S) \subset \bigcap G_m$. Let $y \in \bigcap G_m$. For each $m \geq 1$, there exists an open ball U_m containing y with a centre at $h(x_m)$ for some $x_m \in S$ such that the diameters of U_m and $V_m = h^{-1}(U_m)$ are less than $1/m$. Since $h(x_m) \to y, y \in \overline{h(S)}$. So for any $k, m \geq 1, U_k \bigcap U_m$ contains a $y' \in h(S)$. Then $x' = h^{-1}(y') \in V_k \bigcap V_m$ and

$$d(x_k, x_m) \leq d(x_k, x') + d(x', x_m) < (1/k) + (1/m).$$

Thus $\{x_m\}$ is Cauchy and has a limit x. Then $h(x) = y$ and $y \in h(S) = \bigcap G_m$ which is G_δ.

For the second part of the theorem, it suffices to prove that $[0, 1)^\infty$ is measurably isomorphic to [0,1). Construct $f : [0, 1) \to [0, 1)^\infty$ as follows:

Write $x \in [0,1)$ as $.x_1x_2x_3\cdots =$ its unique binary expansion containing infinitely many zeros. Consider the array

1

2 3

4 5 6

. . . .

Let $f(x) = (f_1(x), f_2(x), \cdots)$ where,

$$f_1(x) = .x_1x_2x_4x_7\cdots, f_2(x) = .x_3x_5x_8\cdots, \text{and so on.}$$

(Go down the columns of the array). Conversely, let $x = (x^1, x^2, \cdots) \in [0,1)^\infty$ with $x^k = .x_1^kx_2^k\cdots$ being the unique binary expansion of x^k containing infinitely many zeros. Let $g(x) =$ the binary decimal whose $n-th$ entry is x_j^k if n appears in the $k-th$ column of the above array j numbers down. It is easy to check (Exercise 1.1) that f, g are measurable and $f = g^{-1}$. □

Remark One can in fact show that given two Borel subsets of two possibly distinct Polish spaces, if they have the same cardinality, they are measurably isomorphic. (See [38, p. 14].)

The closure $\overline{h(S)}$ of $h(S)$ in $[0,1]^\infty$ is compact and will be called the compactification of S.

Let $X_\alpha, \alpha \in I$, be random variables on $(\Omega, \mathcal{F}, P)$ taking values in Polish spaces $S_\alpha, \alpha \in I$, respectively. Then their finite dimensional marginals (i.e., the laws of finite subcollections of $\{X_\alpha\}$) are consistent in the following sense: If $I_1 = \{\alpha_1, \cdots, \alpha_m\} \subset I_2 = \{\beta_1, \cdots, \beta_k\} \subset I$ and μ_1, μ_2 are the respective laws of $(X_{\alpha_1}, \cdots, X_{\alpha_m}), (X_{\beta_1}, \cdots, X_{\beta_k})$, then μ_1 is the image of μ_2 under the projection $\prod_{I_2} S_\alpha \to \prod_{I_1} S_\alpha$. The next theorem shows that this consistency condition is sufficient for the ("canonical") realizability
of such random variables. We shall need the following lemma which holds true for arbitrary (i.e., not necessarily Polish space valued) random variables.

Lemma 1.1.1 $\sigma(X_\alpha, \alpha \in I) = \bigcup \sigma(X_\alpha, \alpha \in J)$ *where the union is over all countable* $J \subset I$.

Proof The right hand side is clearly contained in the left. The claim follows on verifying that the former is a $\sigma-$field (Exercise 1.2). □

Theorem 1.1.2 *(Kolmogorov extension theorem) Let* $S_\alpha, \alpha \in I$, *be Polish spaces and* $\mu_J, J \subset I$ *finite, a consistent family of probability measures on* $S_J = \prod_J S_\alpha$ *respectively. Then there exists a unique probability measure* μ *on* $S_I = \prod_I S_\alpha$ *such that* μ_J *is the image of* μ *under the projection* $S_I \to S_J$.

Proof To start with, let I be countable and S_α compact for all α. Call an $f \in C(S_I)$ a cylindrical function if it is of the form $f(x) = g(x_{\alpha_1}, \cdots, x_{\alpha_n})$ for some $n \geq 1, J = \{\alpha_1, \alpha_2, \cdots, \alpha_n\} \subset I$, $g \in C(S_J)$, with x_α being the $\alpha - th$ component of $x \in S_I$. By the Stone-Weirstrass theorem, the set D of cylindrical functions is dense in $C(S_I)$. The map $F : f \in D \to \int g d\mu_J \in \boldsymbol{R}$ for f, g as above is a bounded linear functional on D satisfying (i) $F(f) \geq 0$ when $f \geq 0$ and (ii) $F(\underline{1}) = 1$ where $\underline{1}$ = the constant function identically equal to one. F uniquely extends to a bounded linear $\overline{F} : C(S_I) \to \boldsymbol{R}$ satisfying (i), (ii). By the Riesz theorem, $\overline{F}(f) = \int f d\mu$ for a probability measure μ on S_I. This μ clearly meets the requirements. Also, it is unique since any two candidates for μ must agree on D and hence on $C(S_I)$.

Next consider the case when $\{S_\alpha\}$ are not necessarily compact. Identify S_α with $h(S_\alpha) \subset [0,1]^\infty$ for h as in Theorem 1.1.1 and let $\overline{S}_\alpha$ denote its compactification $\overline{h(S_\alpha)}$. Let $\overline{S}_J = \prod_J \overline{S}_\alpha$ for $J \subset I$. For $J \subset I$ finite, let $\overline{\mu}_J$ be the probability measure on $\overline{S}_J$ that uniquely extends μ_J. Then $\overline{\mu}_J(\overline{S}_J \backslash S_J) = 0$ for all J and $\{\overline{\mu}_J\}$ are consistent. By the foregoing, there is a unique probability measure $\overline{\mu}$ on $\overline{S}_I$ that projects to $\overline{\mu}_J$ under the projection $\overline{S}_I \to \overline{S}_J$, for each finite $J \subset I$. Since $\overline{\mu}_J(\overline{S}_J \backslash S_J) = 0$ for all finite $J \subset I$, it follows that $\overline{\mu}(\overline{S}_I \backslash S_I) = 0$ (Exercise 1.3). Thus $\overline{\mu}$ restricts to a probability measure μ on S_I satisfying the requirements. Uniqueness of μ is easy to prove.

Finally, for not necessarily countable I, the claim can be proved using the above in conjunction with Lemma 1.1.1 (Exercise 1.4). □

An important consequence of Lemma 1.1.1 is that $\sigma(X_\alpha, \alpha \in I)$ contains only "countably described" sets, that is, sets that can be described in terms of $X_\alpha, \alpha \in J$, for some countable $J \subset I$ (Exercise 1.5). Thus if $S_\alpha = \boldsymbol{R}$ for all $\alpha \in I = [0,1]$, the sets $\{\omega \mid sup \mid X_t(\omega) \mid = 1\}$, $\{\omega \mid$ the map $t \to X_t(\omega)$ is continuous on $[0,1]\}$ may not be in $\sigma(X_t, t \in I)$. We return to this issue in Chapter VI.

Call a probability measure μ on a topological space S tight if for each $\epsilon > 0$, there is a compact $K_\epsilon \subset S$ with $\mu(K_\epsilon) > 1 - \epsilon$. A family $\{\mu_\alpha, \alpha \in I\}$ of probability measures on S is said to be tight if for each $\epsilon > 0$, there is a compact $K_\epsilon \subset S$ such that $\mu_\alpha(K_\epsilon) > 1 - \epsilon$ for all $\alpha \in I$.

Theorem 1.1.3 *(Oxtoby, Ulam) Any probability measure μ on a Polish space S is tight.*

Proof For each $n \geq 1$, let $\{A_{ni}\}$ be open $(1/n)$-balls covering S and $i_n \geq 1$ such that for a given $\epsilon > 0$,

$$\mu(\bigcup_{i \leq i_n} A_{ni}) > 1 - \epsilon 2^{-n-1}, n \geq 1.$$

Then the set $\bigcap_{n\geq 1}\bigcup_{i\leq i_n} A_{ni}$ is totally bounded and thus has a compact closure K satisfying $\mu(K) > 1-\epsilon$. $\square$

Our final result of this section characterizes random variables measurable with respect to $\sigma(X)$. But first we need the following technical lemma. Call a random variable elementary (resp. simple) if it takes at most countably (resp. finitely) many values.

Lemma 1.1.2 *Let S be a Polish space with a complete metric d and X, $\{X_m\}$ S-valued random variables on a common probability space.*

(a) X is the uniform limit (resp. limit) of a sequence of elementary (resp. simple) random variables.

(b) The set $\{\omega \mid \lim_{n\to\infty} X_n(\omega)$ exists$\}$ is measurable.

Proof (a) Let $\{s_n\} \subset S$ be countably dense and B_{nk} the open $(1/k)$–ball centred at s_n. Let $C_{nk} = B_{nk}\backslash\bigcup_{j=1}^{n-1} B_{jk}$ and $\overline{X}_k = s_n$ on $X^{-1}(C_{nk})$, $n \geq 1$. Then $\sup_s d(\overline{X}_k, X) < 1/k$, proving the first claim. For the second claim, define $\{X'_{km}\}$ by $X'_{km} = s_n$ on $X^{-1}(C_{nk})$, $n \leq m$, and $= s_1$ elsewhere. Then $X'_{km} \to \overline{X}_k$ as $m \to \infty$ pointwise for each k. In fact, $X'_{km} = \overline{X}_k$ on $\bigcup_{n=1}^{m} X^{-1}(C_{nk})$. The rest is easy.

(b) The set equals $\{\omega \mid \{X_n(\omega)\}$ is Cauchy$\}$, that is,

$$\bigcap_{k=1}^{\infty}\bigcup_{j}\bigcap_{n\geq j}\bigcap_{m}\{\omega \mid d(X_n(\omega), X_{n+m}(\omega)) < 1/k\}.$$

$\square$

Remark If $S = \boldsymbol{R}$, one may replace "limit" by "monotone limit" in (a) above (Exercise 1.6).

Theorem 1.1.4 *Let $X : (\Omega, \mathcal{F}, P) \to (E, \xi)$ be a random variable and Y another random variable on $(\Omega, \mathcal{F}, P)$ taking values in a Polish space S. Then Y is $\sigma(X)$-measurable if and only if $Y = h(X)$ for a measurable $h : (E, \xi) \to S$.*

Proof Sufficiency is clear. To prove necessity, assume for the time being that Y takes only countably many values $\{b_n\}$. Then $A_n = \{Y = b_n\} \in \sigma(X)$ and therefore are of the form $X^{-1}(B_n)$ for some $B_n \in \xi$ for all n. Let $C_n = B_n\backslash(\bigcup_{j<n} B_j)$. Then $C_n \in \xi$ for all n and $\{C_n\}$ is a disjoint family. Also, $X^{-1}(C_n) = A_n\backslash(\bigcup_{j<n} A_j) = A_n$. Define $h : E \to S$ by: $h = b_n$ on $C_n, n \geq 1$, and $= b_1$ elsewhere. Then h is measurable and $Y = h(X)$. For general (i.e., not necessarily countable valued) Y, let $\{Y_n\}$ be $\sigma(X)$-measurable elementary random variables converging uniformly to Y and let

$Y_n = h_n(X)$ for suitably defined measurable maps $h_n : (E, \xi) \to S, n \geq 1$. Let $h = lim\ h_n$ on $\{lim\ h_n \text{ exists}\}$ and equal to an arbitrary element of S elsewhere. In view of part (b) of the above lemma, h is measurable. Also, $Y = h(X)$. □

Along the lines of the remark following Theorem 1.1.2, one may combine the above with Lemma 1.1.1 to deduce that any Polish space-valued random variable measurable with respect to $\sigma(X_\alpha, \alpha \in I)$ is in fact a measurable function of $X_\alpha, \alpha \in J$, for some countable $J \subset I$.

1.2 Monotone Class Theorems

Monotone class theorems is the generic name for a class of results wherein one extends some property from a family of sets, functions, etc. to a larger family under hypotheses that include a condition to the effect that some property be preserved under an appropriate "monotone limit". We shall give one set theoretic and one functional instance of these. (See [11, Ch. I], for more.) Typical applications are given in the exercises.

A collection Q of subsets of Ω is called a π-system if it is closed under finite intersections and a λ-system if it contains Ω and satisfies

(i) $A, B \in Q, A \subset B$ imply $B - A \in Q$,

(ii) $A_1 \subset A_2 \subset A_3 \subset \cdots$ with $A_i \in Q$ for all i implies $\bigcup_i A_i \in Q$.

If Q is both a π-system and a λ-system, it is a σ-field and vice versa (Exercise 1.7). Arbitrary intersections of λ-systems are seen to be λ-systems and thus it makes sense to speak of $\lambda(\mathcal{B})$ = the smallest λ-systems containing $\mathcal{B}$ for any family $\mathcal{B}$ of subsets of Ω. Denote by $\sigma(\mathcal{B})$ the smallest σ-field containing $\mathcal{B}$. Our "set theoretic" monotone class theorem is the following:

Theorem 1.2.1 *If $\mathcal{B}$ is a π-system, $\lambda(\mathcal{B}) = \sigma(\mathcal{B})$.*

Proof Let $D_1 = \{B \in \lambda(\mathcal{B}) \mid B \bigcap A \in \lambda(\mathcal{B}) \text{ for } A \in \mathcal{B}\}$. D_1 is seen to be a λ-system containing $\mathcal{B}$ and therefore equals $\lambda(\mathcal{B})$. Let $D_2 = \{B \in \lambda(\mathcal{B}) \mid B \bigcap A \in \lambda(\mathcal{B}) \text{ for } A \in \lambda(\mathcal{B})\}$. Again one verifies that D_2 is a λ-system. If $A \in \mathcal{B}, B \bigcap A \in \lambda(\mathcal{B})$ for $B \in D_1 = \lambda(\mathcal{B})$, implying $\mathcal{B} \subset D_2$. Thus $\lambda(\mathcal{B}) = D_2$ and is a π-system and hence a σ-field. Since $\lambda(\mathcal{B}) \subset \sigma(\mathcal{B})$ in any case, they are equal. □

The "monotone" condition here is the (ii) in the definition of a λ-system. The functional monotone class theorem follows next.

Theorem 1.2.2 *Let H be a linear space of bounded real functions on Ω containing constants and closed under both uniform convergence and monotone convergence of uniformly bounded nonnegative functions in H.*

Let $G \subset H$ be closed under multiplication. Then H contains all bounded $\sigma(G)-$measurable functions.

Proof Without any loss of generality, let G contain the constants. By Zorn's lemma, the family of algebras containing G and contained in H, partially ordered by inclusion, contains a maximal element Q. Closure of Q under uniform convergence would be another such algebra and hence by maximality of Q, coincides with it. An analogous argument shows that Q is closed under monotone convergence of uniformly bounded nonnegative elements of itself. Let $\mathcal{B}$ be the collection of subsets whose indicators are in Q. Since Q is an algebra, $\mathcal{B}$ is the π-system. Also, Q contains constants (and is therefore closed under complementation) and is closed under bounded monotone convergence of uniformly bounded nonnegative elements therein. Thus $\mathcal{B}$ is a λ-system and hence a σ-field. Since Q is closed under uniform limits, it follows from Lemma 1.1.2(a) that Q contains all bounded $\mathcal{B}$-measurable functions. So it suffices to show that $\sigma(G) \subset \mathcal{B}$. Let $a > 0$. The map $x \to ((x \bigwedge a) \bigvee 0)/a$ is uniformly approximated by polynomials on compact intervals of $\boldsymbol{R}$. Thus Q is closed under the operation $f \to ((f \bigwedge a) \bigvee 0)/a$. But $I\{f \geq a\}$ for $f \in Q$ is the decreasing limit of $[((f \bigwedge a) \bigvee 0)/a]^n$ as $n \to \infty$ and therefore is in Q. Thus $\{f \geq a\} \in \mathcal{B}$. A similar argument proves this for all $a \in \boldsymbol{R}$. Thus $\sigma(G) \subset \mathcal{B}$. □

1.3 Expectations and Uniform Integrability

For an integrable real random variable X (i.e., $X \in L_1(\Omega, \mathcal{F}, P)$), $E[X] = \int X dP$ is called the mean, expected value or expectation of X. Furthermore, $E[X^m], E[| X |^m], E[(X - E[X])^m]$ for $m \geq 1$ are called respectively the m-th moment, the m-th absolute moment and the m-th centred moment, whenever these exist. The second centred moment is called the variance of X, denoted $var(X)$, and it or its square root (called the standard deviation of X) is often used as a quantitative measure of the spread of $\mathcal{L}(X)$. The following theorem lists some simple but useful consequences of this definition.

Theorem 1.3.1 *(a) (Change of variables formula) Let $X : (\Omega, \mathcal{F}, P) \to (E, \xi)$ be a random variable with law μ and $f : (E, \xi) \to \boldsymbol{R}$ a measurable map such that $f \circ X$ is integrable. Then*

$$\int f(X)dP = \int f(x)d\mu(x).$$

(b) (Jensen's inequality) If $E = \boldsymbol{R}^n$ and $f : \boldsymbol{R}^n \to \boldsymbol{R}$ convex,

$$E[f(X)] \geq f(E[X])$$

whenever both sides are defined.
(c) (Chebyshev's inequality) For $E = \boldsymbol{R}$ *and* $f \geq 0$ *in (a),*

$$E[f(X)] \geq \epsilon P(f(X) \geq \epsilon), \epsilon > 0.$$

Proof (a) In the set-up of Theorem 1.2.2, let $H =$ the set of those f for which the desired equality holds and $G =$ the set of indicator functions. The claim is obvious for $f \in G$. Now use Theorem 1.2.2.
(b) The claim holds with equality for affine f. For convex f, one can write $f(x) = sup_{\alpha \in I} f_\alpha(x)$ for a family of affine functions $f_\alpha, \alpha \in I$ (viz., those whose graphs lie below that of f). Thus

$$E[f(X)] = E[\sup_I f_\alpha(X)] \geq \sup_I E[f_\alpha(X)] = \sup_I f_\alpha(E[X]) = f(E[X]).$$

(c)

$$\begin{aligned} E[f(X)] &= E[f(X)I\{f(X) \geq \epsilon\}] + E[f(X)I\{f(X) < \epsilon\}] \\ &\geq \epsilon P(f(X) \geq \epsilon), \end{aligned}$$

which is the desired inequality. □

A family $\{X_\alpha, \alpha \in I\}$ of real random variables is said to be uniformly integrable (u.i.) if it is integrable and satisfies

$$\lim_{a \to \infty} \sup_\alpha E[| X_\alpha | I\{| X_\alpha | \geq a\}] = 0.$$

The following result provides another equivalent definition.

Theorem 1.3.2 $\{X_\alpha, \alpha \in I\}$ *is u.i. if and only if* $\sup_\alpha E[| X_\alpha |] < \infty$ *and*

$$\lim_{P(A) \to 0} \sup_\alpha \int_A | X_\alpha | dP = 0.$$

Proof Assuming the latter conditions, let $\epsilon, \delta > 0$. For sufficiently large a,

$$P(| X_\alpha | \geq a) \leq E[| X_\alpha |]/a \leq \sup_a E[| X_\alpha |]/a < \delta, \ \alpha \in I.$$

Picking δ sufficiently small, one then has

$$\sup_{\alpha \in I} \int_{\{|X_\beta| \geq a\}} | X_\alpha | dP < \epsilon, \ \beta \in I.$$

Thus

$$\int_{\{|X_\beta|\geq a\}} | X_\beta | \, dP < \epsilon, \beta \in I,$$

and $\{X_\alpha, \alpha \in I\}$ are u.i. Conversely, let $\{X_\alpha\}$ be u.i., $\epsilon > 0$. For sufficiently large a,

$$\sup_I \int_{\{|X_\alpha|\geq a\}} | X_\alpha | \, dP < \epsilon.$$

Thus $\sup_I E[| X_\alpha |] < a + \epsilon < \infty$. Let $\delta = \epsilon/a$ and A an event such that $P(A) < \delta$. Then for all $\alpha \in I$,

$$\begin{aligned} \int_A | X_\alpha | \, dP &= \int_{A\bigcap\{|X_\alpha|\geq a\}} | X_\alpha | \, dP + \int_{A\bigcap\{|X_\alpha|<a\}} | X_\alpha | \, dP \\ &\leq \int_{\{|X_\alpha|\geq a\}} | X_\alpha | \, dP + aP(A) < 2\epsilon. \end{aligned}$$

The claim follows. □

Corollary 1.3.1 *Let $X_n, n = 1, 2, \cdots, \infty$, be real random variables on $(\Omega, \mathcal{F}, P)$ with $P(X_n \to X_\infty) = 1$. If $X_n, n = 1, 2, \cdots$, are u.i., then $X_n \to X_\infty$ in $L_1(\Omega, \mathcal{F}, P)$.*

Proof By Fatou's lemma, X_∞ is integrable. If $X_n, n = 1, 2, \cdots$, are u.i., then so are $X_n, n = 1, 2, \cdots, \infty$. For any $\epsilon > 0$,

$$\begin{aligned} E[| X_n - X_\infty |] &\leq \epsilon + E[| X_n - X_\infty | I\{| X_n - X_\infty | \geq \epsilon\}] \\ &\leq \epsilon + 2 \sup_{1\leq j\leq\infty} \int_{\{|X_n - X_\infty|\geq\epsilon\}} | X_j | \, dP. \end{aligned}$$

Since $P(| X_n - X_\infty | \geq \epsilon) \to 0$ as $n \to \infty$, the right hand side is less than 2ϵ for sufficiently large n. The claim follows. □

Note that $X_n \to X_\infty$ in $L_1(\Omega, \mathcal{F}, P)$ automatically implies that $X_n, n = 1, 2, \cdots$, are u.i. (Exercise 1.8). A related result is:

Theorem 1.3.3 *Let $X_n, n = 1, 2, \cdots, \infty$ be integrable random variables on $(\Omega, \mathcal{F}, P)$ with $X_n \geq 0$ a.s. for all n and $P(X_n \to X_\infty) = 1$. Then $X_n \to X_\infty$ in $L_1(\Omega, \mathcal{F}, P)$ if and only if $E[X_n] \to E[X_\infty]$.*

Proof Necessity is clear. To prove sufficiency, observe that

$$0 \leq (X_\infty - X_n)^+ \leq X_\infty$$

and thus $E[(X_\infty - X_n)^+] \to 0$ by the dominated convergence theorem. Then

$$E[| X_\infty - X_n |] = 2E[(X_\infty - X_n)^+] - E[X_\infty - X_n] \to 0.$$

□

The next result gives a useful characterization of uniform integrability.

Theorem 1.3.4 *(de la Vallée-Poussin) $H \subset L_1(\Omega, \mathcal{F}, P)$ is u.i. if and only if there exists a measurable $G : \boldsymbol{R}^+ \to \boldsymbol{R}^+$ such that*

$$\lim_{t\to\infty} G(t)/t = \infty \text{ and } \sup_{X\in H} E[G(|X|)] < \infty.$$

Proof Assuming that the latter conditions hold, let $\delta > 0$ and $a = M/\delta$ where $M = \sup_H E[G(\mid X \mid)]$. Pick $c > 0$ such that $G(t)/t \geq a$ for $t \geq c$. Then $\mid X \mid \leq G(\mid X \mid)/a$ on $\{\mid X \mid \geq c\}$ and

$$\int_{\{|X|\geq c\}} \mid X \mid dP \leq \frac{1}{a}\int_{\{|X|\geq c\}} G(\mid X \mid)dP \leq \frac{M}{a} = \delta$$

for $X \in H$. Also, $E[\mid X \mid] \leq c+\delta$. Thus H is u.i. Conversely, let $X \in H$ and $a_n(X) = P(\mid X \mid \geq n)$. Let $g : \boldsymbol{R}^+ \to \boldsymbol{R}^+$ be an increasing function taking a constant value g_n on $[n, n+1), n \geq 0$, with $g_0 = 0$. Let $G(t) = \int_0^t g(s)ds$. Since $g_0 = 0$,

$$\begin{aligned} E[G(\mid X \mid)] &\leq g_1 P(1 \leq \mid X \mid < 2) + (g_1 + g_2)P(2 \leq \mid X \mid < 3) + \cdots \\ &= \sum_{n=1}^{\infty} g_n a_n(X). \end{aligned}$$

Pick integers $\{c_n\}$ increasing to ∞ such that

$$\int_{\{|X|\geq c_n\}} \mid X \mid dP \leq 2^{-n} \text{ for } X \in H.$$

Then

$$2^{-n} \geq \int_{\{|X|\geq c_n\}} \mid X \mid dP \geq \sum_{k=c_n}^{\infty} kP(k \leq \mid X \mid < k+1) \geq \sum_{m=c_n}^{\infty} a_m(X).$$

Thus

$$\sum_{n=1}^{\infty} \sum_{m=c_n}^{\infty} a_m(X) < \infty$$

uniformly in $X \in H$. Let $g_m =$ the number of integers n such that $c_n \leq m, m = 1, 2, \cdots$. Then the above double summation equals $\sum_n g_n a_n(X)$. By the foregoing, the G corresponding to this choice of $\{g_m\}$ satisfies our requirements. □

The "if" part with $G(t) = t^\alpha, \alpha > 1$, gives a useful test of uniform integrability in practice. Our last result of this section, stated here without proof, links uniform integrability with weak compactness in $L_1(\Omega, \mathcal{F}, P)$.

Theorem 1.3.5 *(Dunford–Pettis compactness criterion) A subset H of $L_1(\Omega, \mathcal{F}, P)$ is u.i. if and only if it is relatively weakly compact which, in turn, is true if and only if each sequence in H has a weakly convergent subsequence.*

See [11, pp. 27–II], for a proof. Note that the second equivalence in the statement of the theorem is not trivial because the weak topology of $L_1(\Omega, \mathcal{F}, P)$ need not be metrizable.

1.4 Independence

It makes intuitive sense to call two events independent if the occurrence of one does not affect the likelihood of the other and the probability of their joint occurrence is simply the product of their individual probabilities. (Think of the relative frequencies in repeated trials.) Thus independence of $A, B \in \mathcal{F}$ is formally defined by the condition $P(AB) = P(A)P(B)$. (We follow a standard convention in probability theory in abbreviating $A \bigcap B$ to AB.) One observes that if A, B are independent, so are A, B^c or A^c, B or A^c, B^c (Exercise 1.9). Also, Ω and ϕ are independent of all events. Thus any set from the sub-σ-field $\{\phi, \Omega, A, A^c\}$ is independent of any set from the sub-σ-field $\{\phi, \Omega, B, B^c\}$. More generally, two sub-σ-fields $\mathcal{F}_1, \mathcal{F}_2$ of $\mathcal{F}$ are said to be independent if every event from one is independent of any event from the other. Two random variables X, Y on $(\Omega, \mathcal{F}, P)$ are said to be independent if $\sigma(X), \sigma(Y)$ are. This is equivalent to saying that $\mathcal{L}((X,Y)) = \mathcal{L}(X) \times \mathcal{L}(Y)$, that is, a product measure. (Alternatively, X, Y are independent if $E[f(X)g(Y)] = E[f(X)]E[g(Y)]$ for all bounded measurable f, g.) Thus the independence of two events is synonymous with that of their indicators. These definitions extend in an obvious manner to finite collections of events, σ-fields or random variables. An arbitrary family of events, σ-fields or random variables is said to be independent if every finite subfamily thereof is. One can also concoct mixed definitions, such as the independence of an event and a σ-field, in an obvious manner. Finally, it is easy to see that if $X_\alpha, \alpha \in I$, is a family of independent random variables, so is $f_\alpha(X_\alpha), \alpha \in I$, for any measurable maps f_α into some measurable spaces that may depend on α (Exercise 1.10).

Remarks (i) A family $\{X_\alpha, \alpha \in I\}$ of random variables is said to be pairwise independent if any two members of this family are independent of each other. Clearly, independence implies pairwise independence. The following example shows that the converse is false: Let X, Y be independent identically distributed (i.i.d.) random variables taking values ± 1, with $P(X = 1) = P(X = -1) = \frac{1}{2}$. Let $Z = XY$. Then it is easily verified that

(X, Y, Z) are pairwise independent but not independent (Exercise 1.11).
(ii) The definition of independence may be motivated as follows (after Kac and Ulam): Intuition suggests that if A, B are independent events, $P(AB)$ should depend on only $P(A)$ and $P(B)$. Thus $P(AB) = f(P(A), P(B))$ for some $f : [0,1] \times [0,1] \to [0,1]$. Clearly f is symmetric (i.e., $f(x,y) = f(y,x)$). It is reasonable to suppose that f is continuous. (Why?) Also, if A is independent of $B_1, B_2, \cdots$ and B_i's are disjoint, it makes sense to demand that $f(P(A), P(\bigcup B_i)) = f(P(A), \sum_i P(B_i)) = \sum_i f(P(A), P(B_i))$. These conditions together imply that $f(x,y) = xy$ (Exercise 1.12).

The notion of independence plays a major role in probability theory. Here we give two simple consequences of the concept. Let $X_1, X_2, \cdots$ be a sequence of random variables on $(\Omega, \mathcal{F}, P)$. Then $\tau = \bigcap_n \sigma(X_n, X_{n+1}, \cdots)$ is called the tail σ-field of $\{X_n\}$.

Theorem 1.4.1 *(Kolmogorov's zero-one law) If $\{X_n\}$ are independent, τ is trivial (i.e., $A \in \tau$ implies $P(A) = 0$ or 1).*

Proof Let $A \in \tau \subset \sigma(X_n, X_{n+1}, \cdots)$, $n \geq 1$. The joint law of $(X_1, \cdots, X_n)$ and $(X_{n+1}, X_{n+2}, \cdots)$ is a product measure for each n. Thus A is independent of $(X_1, \cdots, X_n)$ for each n and hence of $(X_1, X_2, \cdots)$. But $A \in \sigma(X_1, X_2, \cdots)$. Thus A is independent of itself, implying $P(A) = P(A)^2$, that is, $P(A) = 0$ or 1. □

If $\{A_n\}$ are events in $\mathcal{F}$, the event $\{A_n \text{ i.o.}\}$ (i.o. for "infinitely often") stands for $\{\omega \in \Omega \mid \omega \in A_n \text{ for infinitely many } n\} = \bigcap_{n=1}^{\infty} \bigcup_{m=n}^{\infty} A_m$.

Theorem 1.4.2 *(Borel–Cantelli lemma) Let $\{A_n\}$ be as above.*
(i) If $\sum P(A_n) < \infty$, then $P(A_n \text{ i.o.}) = 0$.
(ii) If $\sum P(A_n) = \infty$ and $\{A_n\}$ are independent, then $P(A_n \text{ i.o.}) = 1$.

Proof (i) By the monotone convergence theorem,

$$\infty > \sum_n P(A_n) = E[\sum_n I_{A_n}].$$

Thus $\sum_n I_{A_n} < \infty$ a.s.
(ii) It suffices to show that $exp(-\sum_n I_{A_n}) = 0$ a.s. Equivalently, one may show that $E[exp(-\sum_n I_{A_n})] = 0$. But

$$\begin{aligned} E\left[exp\left(-\sum_{n=1}^{N} I_{A_n}\right)\right] &= \prod_{n=1}^{N} E\left[exp\left(-I_{A_n}\right)\right] \\ &= \prod_{n=1}^{N} \left[e^{-1}P(A_n) + (1 - P(A_n))\right] \\ &= \prod_{n=1}^{N} (1 - (1 - e^{-1})P(A_n)) \end{aligned}$$

$$\leq \prod_{n=1}^{N} exp(-(1-e^{-1})P(A_n))$$
$$= exp(-(1-e^{-1})\sum_{n=1}^{N} P(A_n)),$$

where we have used the inequality $1 - x \leq e^{-x}$. Let $N \to \infty$ to conclude. □

1.5 Convergence Concepts

The usual convergence concepts of measure theory carry over to probability theory with a slightly altered terminology. Thus given a Polish space S and S-valued random variables $X_1, X_2, \cdots, X_\infty$ on $(\Omega, \mathcal{F}, P)$, we say that $X_n \to X_\infty$ a.s. (almost surely) if $P(X_n \to X_\infty) = 1$ and in probability if $P(d(X_n, X_\infty) \geq \epsilon) \to 0$ for all $\epsilon > 0$. The former clearly implies the latter which, incidentally, is dictated only by the joint laws of pairs $(X_n, X_\infty), n = 1, 2, \cdots$. For $S = \mathbf{R}, X_n \to X_\infty$ in p-th mean if it does so in $L_p(\Omega, \mathcal{F}, P), 1 \leq p < \infty$. It is easy to see that p-th mean convergence implies convergence in probability (Exercise 1.13). The converse is false in general unless additional hypotheses are made. For example, convergence of $\{X_n\}$ to X_∞ in probability implies their convergence in mean (i.e., in $L_1(\Omega, \mathcal{F}, P)$) if $\{X_n\}$ are u.i. This is proved exactly as in Corollary 1.3.1. The convergence in probability can be metrized as described below. We assume throughout that d is a complete metric on S and takes values in [0,1].

Theorem 1.5.1 *The S-valued random variables on $(\Omega, \mathcal{F}, P)$ form a complete metric space under the metric $\rho(X, Y) = E[d(X, Y)]$. The convergence in this metric is equivalent to convergence in probability.*

Proof It is easy to check that ρ is a metric (Exercise 1.14). For $\epsilon > 0$, one has

$$\epsilon + P(d(X, Y) \geq \epsilon) \geq E[d(X, Y)] \geq \epsilon P(d(X, Y) \geq \epsilon).$$

Thus ρ-convergence is equivalent to the convergence in probability. Also, a sequence $\{X_n\}$ of S-valued random variables is Cauchy with respect to ρ if and only if it is "Cauchy in probability", that is,

$$\lim_{n,m\to\infty} P(d(X_n, X_m) \geq \epsilon) = 0 \text{ for } \epsilon > 0.$$

Thus we may pick integers $k_n \uparrow \infty$ such that

$$P(d(X_{k_n}, X_{k_n+m}) > 2^{-n}) < 2^{-n}, n, m \geq 1,$$

leading to

$$\sum_{n=1}^{\infty} P(d(X_{k_n}, X_{k_{n+1}}) > 2^{-n}) < \infty.$$

By the Borel–Cantelli lemma, the sequence $\{d(X_{k_n}, X_{k_{n+1}})\}$ is eventually dominated by the sequence $\{2^{-n}\}$ a.s. Hence $\{X_{k_n}\}$ is a.s. Cauchy with respect to d and therefore converges a.s. to a random variable Y. Then it does so also in probability and in metric ρ. Since $\{X_n\}$ is Cauchy in ρ, it must converge to Y in ρ, implying that ρ is complete. □

Corollary 1.5.1 *$X_n \to Y$ in probability if and only if for each subsequence of $\{X_n\}$, there is a further subsequence which converges to Y a.s.*

Proof The "only if" part can be proved by using Borel–Cantelli lemma as in the proof of the above theorem. Since a.s. convergence implies convergence in probability, the "if" part follows from the fact that the convergence in probability is a metric convergence. □

Corollary 1.5.2 *Suppose that convergence in probability does not imply a.s. convergence on $(\Omega, \mathcal{F}, P)$. Then it is impossible to find a metric δ on the space of random variables on $(\Omega, \mathcal{F}, P)$ such that $\delta(X_n, X) \to 0$ if and only if $X_n \to X$ a.s.*

Proof Suppose it is possible to find such a δ. Let $X_n, n = 1, 2, \cdots, \infty$ be random variables on $(\Omega, \mathcal{F}, P)$ such that $X_n \to X_\infty$ in probability but not a.s. Then $\delta(X_{n(k)}, X_\infty) \geq \epsilon > 0$ for some ϵ and some $\{n(k)\} \subset \{n\}$. By the preceding corollary, there is a subsequence $\{k(m)\}$ of $\{k\}$ such that $X_{n(k(m))} \to X_\infty$ a.s. Then $\delta(X_{n(k(m))}, X_\infty) \to 0$, a contradiction. □

We shall now explore the conditions under which a.s. convergence and convergence in probability are equivalent. A set $A \in \mathcal{F}$ is said to be a P-atom if for any $B \in \mathcal{F}$, either $P(B \bigcap A) = P(A)$ or $P(B \bigcap A) = 0$ (i.e., modulo a set of zero P-measure, either B or B^c contains A.) If A_1, A_2 are P-atoms, they are either essentially identical (i.e., $P(A_1 \Delta A_2) = 0$) or essentially disjoint (i.e., $P(A_1 \bigcap A_2) = 0$). It is clear that there can be, at most, countably many essentially disjoint P-atoms with strictly positive probability. P is said to be purely atomic if there is a countable family $\{A_i\}$ of essentially disjoint P-atoms such that $P(\bigcup A_i) = 1$. It is said to be purely nonatomic if the only P-atoms are the null events. Any probability measure is either purely atomic, purely nonatomic or can be expressed as a strict convex combination of a purely atomic and a purely nonatomic probability measure which are mutually singular (Exercise 1.15).

Theorem 1.5.2 *Convergence in probability and a.s. convergence on $(\Omega, \mathcal{F}, P)$ are equivalent if and only if P is purely atomic.*

Proof Clearly, any random variable on $(\Omega, \mathcal{F}, P)$ will be a.s. constant on each P-atom. Thus if P is purely atomic, each random variable a.s. equals an elementary random variable. Let $X_n \to X_\infty$ in probability. Let $\{A_n\}$ be an enumeration of essentially distinct P-atoms such that $P(A_n) = b_n > 0, n \geq 1, \Sigma b_n = 1$. Let $X_n = a_{nm}$ on A_m for $n = 1, 2, \cdots, \infty, m \geq 1$. If $a_{nm} \not\to a_{\infty m}$ for some m, then $\mid a_{nm} - a_{\infty m} \mid \geq \epsilon$ i.o. for some $\epsilon > 0$ and therefore $P(\mid X_n - X_\infty \mid \geq \epsilon) \geq b_m$ i.o., contradicting the convergence of $\{X_n\}$ to X_∞ in probability. Thus $a_{nm} \to a_{\infty m}$ for all m and therefore $X_n \to X_\infty$ a.s.

For the converse, first consider the case when P is purely nonatomic. We claim that there exist $A_1, A_2 = A_1^c \subset \mathcal{F}$ with $P(A_1) = P(A_2) = \frac{1}{2}$. Suppose not. Define a partial order $\leq$ on $\mathcal{F}$ by: $A \leq B$ if $P(B^c \bigcap A) = 0$. Let $Q = \{B \in \mathcal{F} \mid P(B) \leq \frac{1}{2}\}$. Let $\{B_\alpha, \alpha \in I\} \subset Q$ be totally ordered under $\leq$. Let $a = sup_\alpha P(B_\alpha)$. Pick $\alpha(n), n \geq 1$, in I such that $P(B_{\alpha(n)}) \geq a - (1/n)$ for $n \geq 1$. Then $\bigcup_n B_{\alpha(n)}$ is an upper bound on $\{B_\alpha, \alpha \in I\}$ in Q (Exercise 1.16). By Zorn's lemma, Q has a maximal element D. By hypothesis, $P(D) < \frac{1}{2}$. Let $\mathcal{B} = \{B \in \mathcal{F} \mid D \subset B$ and $P(B) > \frac{1}{2}\}$. Argue as above to deduce that $\mathcal{B}$ has a minimal element D'. By hypothesis, $P(D') > \frac{1}{2}$. Let $G = D' \backslash D$. Then $P(G) > 0$ and hence G is not a P-atom. Thus we can write $G = G_1 \bigcup G_2$ for $G_1, G_2 \in \mathcal{F}$, satisfying $G_1 \bigcap G_2 = \phi$ and $P(G_i) > 0, i = 1, 2$. Then $P(D) < P(D \bigcup G_1) < P(D')$ and $D \bigcup G_1$ violates either the maximality of D in Q or the minimality of D' in $\mathcal{B}$. Thus the aforementioned sets $A_1, A_2 = A_1^c$ must exist. Repeat the argument to split $A_i, i = 1, 2$, into disjoint sets $A_{ij}, j = 1, 2$, such that $P(A_{ij}) = \frac{1}{4}$. In turn, split each $A_{ij}, i, j = 1, 2$, into disjoint sets $A_{ijk}, k = 1, 2$, with $P(A_{ijk}) = \frac{1}{8}$ and so on. Let $X_1 = I_{A_1}, X_2 = I_{A_2}, X_3 = I_{A_{11}}, X_4 = I_{A_{12}}, X_5 = I_{A_{21}}, X_6 = I_{A_{22}}, X_7 = I_{A_{111}}, X_8 = I_{A_{112}}, \cdots$. Then it follows that $P(\mid X_n \mid \geq \epsilon) \to 0$ for $\epsilon > 0$, implying $X_n \to 0$ in probability. But for any $\omega \in \Omega, X_n(\omega) = 1$ i.o. and thus $X_n \not\to 0$ a.s. This settles the issue for purely nonatomic P.

For a general P which is not purely atomic, write $P = \alpha P_1 + (1 - \alpha) P_2$ for some $\alpha \in [0,1)$ and P_1, P_2 mutually singular probability measures which are respectively purely atomic and purely nonatomic. Construct $\{X_n\}$ as above on $(\Omega, \mathcal{F}, P_2)$. Let $C \in \mathcal{F}$ be such that $P_1(C) = 0, P_2(C) = 1$. Define $X'_n = X_n I_C, n \geq 1$. Then $X'_n \to 0$ in probability on $(\Omega, \mathcal{F}, P)$, but not a.s. □

For further reading related to this chapter, see [7, 8, 9, 11, 34, 37, 38].

1.6 Additional Exercises

(1.17) In a measurable space $(\Omega, \mathcal{F})$, atoms are the equivalence classes under the equivalence relation: $\omega \sim \omega'$ if and only if $I_A(\omega) = I_A(\omega')$ for all $A \in \mathcal{F}$.

(a) Show that A is an atom if and only if every measurable $f : (\Omega, \mathcal{F}) \to \boldsymbol{R}$ is constant on A.

(b) Show that if $\mathcal{F}$ is countably generated (i.e., $\mathcal{F}$ is the smallest σ-field containing a prescribed countable collection of sets), then its atoms are measurable.

(1.18) Show that $\mathcal{F}$ above is countably generated if and only if it can be written as $\sigma(X)$ for a measurable X: $(\Omega, \mathcal{F}) \to \boldsymbol{R}$.

(1.19) Show that any probability measure μ on $\boldsymbol{R}$ is the image of the uniform measure on [0,1] under some measurable map f: $[0,1] \to \boldsymbol{R}$. (Hint: Take $f(x) = sup\{y \mid \mu((-\infty, y]) \leq x\}$ for $x \in (0,1)$.)

(1.20) Show that every probability measure on a metric space S equipped with its Borel σ-field satisfies:

$$\begin{aligned} \mu(A) &= inf\{\mu(G) \mid G \text{ open}, A \subset G\} \\ &= sup\{\mu(F) \mid F \text{ closed}, F \subset A\} \end{aligned}$$

for all Borel sets A. (Hint: Show that the sets satisfying this condition form a σ-field containing all closed sets.)

(1.21) Show that a family A of probability measures on $\boldsymbol{R}^n$ is tight if and only if there exists a function $f \in C(\boldsymbol{R}^n)$ satisfying $f(x) \to \infty$ as $\|x\| \to \infty$ such that

$$\sup_{m \in A} \int f(x)\, m(dx) < \infty.$$

(1.22) The following illustrate the use of the monotone class theorems:

(a) Let μ, ν be probability measures on $\boldsymbol{R}^2$ such that $\mu(A) = \nu(A)$ for all finite rectangles A with sides parallel to the axes. Show that $\mu = \nu$.

(b) Let X, Y be real random variables such that $E[f(X)g(Y)] = E[f(X)]E[g(Y)]$ for all smooth compactly supported f, g. Show that X, Y are independent.

(1.23) (a) Let $f, g : \boldsymbol{R} \to \boldsymbol{R}$ be either both nonincreasing or both nondecreasing. Show that

$$(f(x) - f(y))(g(x) - g(y)) \geq 0, \ x, y \in \boldsymbol{R}.$$

(b) Let f, g be as above and X, Y real random variables such that $\mathcal{L}((X,Y)) = \mathcal{L}((Y,X))$. Show that

$$E[f(X)g(X)] \geq E[f(X)g(Y)],$$

assuming that the expectations are defined.

(c) In the above set-up, show that

$$E[f(X)g(X)] \geq E[f(X)]E[g(X)].$$

(This inequality is due to Chebyshev.)

(1.24) Let X, Y, Z be random variables taking values in $[-1, 1]$. Show that

$$1 - E[XY] \geq | \, E[XZ] - E[YZ] \, | \, .$$

(This is called Bell's inequality.)

(1.25) Let X be a nonnegative random variable. Show that

$$E[X^p] = p \int_0^\infty t^{p-1} P(X \geq t) dt, \ p \geq 1.$$

(1.26) Let $\{X_n\}$ be i.i.d. real-valued random variables. Show that $X_n/n \to 0$ a.s. if and only if $E[| \, X_1 \, |] < \infty$.

(1.27) Let $\{X_n\}$ be i.i.d. random variables which are not a.s. a constant. Show that $P(X_n \text{ converges}) = 0$.

(1.28) Give examples of real random variables $\{X_n\}$ which (a) converge a.s. but not in mean, (b) converge in mean but not a.s.

2

Spaces of Probability Measures

2.1 The Prohorov Topology

Let S be a Polish space with a complete metric d taking values in $[0,1]$ and $\boldsymbol{P}(S)$ the space of probability measures on S. Recall the map $h : S \to [0,1]^\infty$ of Theorem 1.1.1. Since $\overline{h(S)}$ is compact, $C(\overline{h(S)})$ is separable. Let $\{\overline{f}_i\}$ be countable dense in the unit ball of $C(\overline{h(S)})$ and $\{f'_i\}$ their restrictions to $h(S)$. Define $\{f_i\} \subset C_b(S)$(= the space of bounded continuous functions $S \to \boldsymbol{R}$) by $f_i = f'_i \circ h, i \geq 1$. Then $\{f_i\}$ is a separating class for $\boldsymbol{P}(S)$, i.e., $\int f_i d\mu = \int f_i d\nu$, $i \geq 1$, for $\mu, \nu \in \boldsymbol{P}(S)$ implies $\mu = \nu$ (Exercise 2.1). Thus

$$\rho(\mu, \nu) = \sum_{n=1}^{\infty} 2^{-n} \mid \int f_n d\mu - \int f_n d\nu \mid$$

defines a metric on $\boldsymbol{P}(S)$. The metric topology of ρ is clearly coarser than that induced by the total variation norm. We call this the Prohorov topology.

Theorem 2.1.1 *The following are equivalent:*

(i) $\rho(\mu_n, \mu) \to 0$,

(ii) $\int f d\mu_n \to \int f d\mu$ *for* $f \in C_b(S)$,

(iii) $\int f d\mu_n \to \int f d\mu$ *for all* f *that are bounded and uniformly continuous with respect to some metric* d' *on* S *equivalent to* d,

(iv) $\limsup_{n\to\infty} \mu_n(F) \le \mu(F)$ *for closed* $F \subset S$,

(v) $\liminf_{n\to\infty} \mu_n(G) \ge \mu(G)$ *for open* $G \subset S$,

(vi) $\lim_{n\to\infty} \mu_n(A) = \mu(A)$ *for Borel* $A \subset S$ *satisfying* $\mu(\partial A) = 0$, *where* ∂A *denotes the boundary of* A.

Proof Clearly (ii) implies (iii). Also, (iv) and (v) are equivalent on identifying G with F^c. Let (iii) hold for some d'. For closed $F \subset S$, $g_n \downarrow I_F$ for $g_n = (1/(1+d'(x,F)))^n, n \ge 1$, which are uniformly continuous with respect to d'. Thus

$$\limsup_{n\to\infty} \mu_n(F) \le \limsup_{n\to\infty} \int g_i d\mu_n = \int g_i d\mu,\ i \ge 1$$

Letting $i \to \infty$ on the right hand side, (iv) follows. (vi) follows from (iv), (v) by applying them respectively to the closure and the interior of a Borel A satisfying $\mu(\partial A) = 0$. Let (vi) hold. Take $f \in C_b(S)$ and for a prescribed $\epsilon > 0$, let $N \ge 1, a_0 < a_1 < a_2 < \cdots < a_N$ be such that for $\| f \| = \sup_x |f(x)|$,

(a) $\| f \| - 1 = a_0 < a_1 < \cdots < a_N = \| f \| + 1$,

(b) $\mu(\{x \mid f(x) = a_i\}) = 0$ for all i, and

(c) $a_i - a_{i-1} \le \epsilon, 1 \le i \le N$.

This is always possible (Exercise 2.2). Let $B_i = \{x \mid a_{i-1} \le f(x) < a_i\}$ for $1 \le i < N$. Then B_i's are disjoint, $\bigcup_i B_i = S$ and $\mu(\partial B_i) = 0$ for all i. Moreover,

$$\| f - \sum_{i=1}^{N} a_i I_{B_i} \| \le \epsilon.$$

Therefore

$$\begin{aligned}
&\limsup_{n\to\infty} \left| \int f d\mu_n - \int f d\mu \right| \\
&\le \ 2\epsilon + \limsup_{n\to\infty} \left| \int \left(\sum_{i=1}^{N} a I_{B_i} \right) d\mu_n - \int \left(\sum_{i=1}^{N} a I_{B_i} \right) d\mu \right| \\
&\le \ 2\epsilon + \sum_{i=1}^{N} | a_i | \left(\limsup_{n\to\infty} | \mu_n(B_i) - \mu(B_i) | \right) \\
&= \ 2\epsilon.
\end{aligned}$$

Since $\epsilon > 0$ was arbitrary, (ii) follows. Thus (ii) - (vi) are equivalent. (ii) clearly implies (i). Now let (i) hold. Let d_1 be a metric on $[0,1]^\infty$ and define a metric d_2 on S by $d_2(x,y) = d_1(h(x),h(y))$ where h is as in Theorem 1.1.1. Since $h : S \to h(S)$ is a homeomorphism, d_2 is a metric equivalent to d. Also, $f \in C_b(S)$ is uniformly continuous with respect to d_2 if and only if $f' = f \circ h^{-1}$ is uniformly continuous on $h(S)$ with respect to d_1. The latter holds if and only if f' is the restriction to $h(S)$ of some $\overline{f} \in C(\overline{h(S)})$. Letting $C_u(S) =$ the space of such f with the supremum norm, $f \to \overline{f}$ is an isometry between $C_u(S)$ and $C(\overline{h(S)})$. Since $\{\overline{f}_i\}$ above are dense in the unit ball of $C(\overline{h(S)})$, $\{f_i\}$ are dense in the unit ball of $C_u(S)$ and thus (i) implies (iii) for $d' = d_2$. This completes the proof. □

Corollary 2.1.1 *Any of the following is a local base at $\mu \in \boldsymbol{P}(S)$ for the Prohorov topology:*

(i) $\{\nu \mid \mid \int f_i d\nu - \int f_i d\mu \mid < \epsilon_i, 1 \leq i \leq k\}, k \geq 1, \epsilon_i > 0, f_i \in C_b(S)$.

(ii) Same as (i) with f_i's bounded uniformly continuous with respect to some metric d' on S equivalent to d.

(iii) $\{\nu \mid \nu(F_i) < \mu(F_i) + \epsilon_i, 1 \leq i \leq k\}, k \geq 1, \epsilon_i > 0, F_i \subset S$ *closed.*

(iv) $\{\nu \mid \nu(G_i) > \mu(G_i) - \epsilon_i, 1 \leq i \leq k\}, k \geq 1, \epsilon_i > 0, G_i \subset S$ *open.*

(v) $\{\nu \mid \mid \nu(A_i) - \mu(A_i) \mid < \epsilon_i, 1 \leq i \leq k\}, k \geq 1, \epsilon_i > 0, A_i \subset S$ *Borel with* $\mu(\partial A_i) = 0$.

The proof follows along similar lines as the preceding theorem (Exercise 2.3).

Corollary 2.1.2 $\boldsymbol{P}(S)$ *is separable.*

Proof Consider a neighbourhood H of μ of type (iii) above. For every nonempty set B_j in the finite partition generated by the collection $\{F_i, 1 \leq i \leq k\}$ featuring in the definition of H, take a point $b(j) \in B_j$. The probability measure $\sum_j \mu(B_j)\delta_{b(j)}$, (where δ_x denotes the Dirac measure at x) is in H. Thus probability measures with finite supports are dense in $\boldsymbol{P}(S)$. Hence the countable set of probability measures whose supports are finite and contained in a prescribed countable dense subset of S and which have a rational mass at each point of the support, is dense in $\boldsymbol{P}(S)$. □

Corollary 2.1.3 *Let $X_n, n = 1, 2, \cdots, \infty$ be S-valued random variables, possibly defined on different probability spaces. Let S' be another Polish space and $h : S \to S'$ continuous. If $X_n \to X_\infty$ in law, then $h(X_n) \to h(X_\infty)$ in law.*

The proof is straightforward from (*ii*) of the above theorem.

A family $\{f_\alpha, \alpha \in I\} \subset C_b(S)$ is called a convergence determining class for $\boldsymbol{P}(S)$ if $\int f_\alpha d\mu_n \to \int f_\alpha d\mu_\infty, \alpha \in I$, for $\mu_n, n = 1, 2, \cdots, \infty$, in $\boldsymbol{P}(S)$, implies $\mu_n \to \mu_\infty$ in $\boldsymbol{P}(S)$. A convergence determining class is clearly a separating class, but the converse need not hold as the following example shows. Let $S = \boldsymbol{R}$ and $\{f_\alpha\}$ = the family of continuous periodic functions $\boldsymbol{R} \to \boldsymbol{R}$ with period = some integer $N, N = 1, 2, \cdots$. This is a separating class (Exercise 2.4). Let μ_n = the Dirac measure at $n!$ for $n = 1, 2, \cdots$, and = the Dirac measure at 0 for $n = 0$. Then for each α, $\int f_\alpha d\mu_n \to \int f_\alpha d\mu_0$. But $\mu_n \not\to \mu_0$ in $\boldsymbol{P}(S)$, implying that $\{f_\alpha\}$ is not a convergence determining class. The functions $\{f_i\}$ in the definition of ρ above form a countable convergence determining class. Other equivalent metrics could be defined by replacing them by other countable convergence determining classes.

In conclusion, an interesting characterization of $C(\boldsymbol{P}(S))$ for compact S due to Dubins follows:

Theorem 2.1.2 *Let $S^\infty = S \times S \times \cdots$, and for $\mu \in \boldsymbol{P}(S)$, $\mu^\infty = \mu \times \mu \times \cdots \in \boldsymbol{P}(S^\infty)$. Then any $F \in C(\boldsymbol{P}(S))$ can be written as $F(\mu) = \int f d\mu^\infty$ for some $f \in C(S^\infty)$.*

See [16] for details.

2.2 Skorohod's Theorem

Let $X_n, n = 1, 2, \cdots, \infty$ be S-valued random variables, not necessarily defined on the same probability space and $\mu_n, n = 1, 2, \cdots, \infty$ their laws. We say that $X_n \to X_\infty$ in law if $\mu_n \to \mu_\infty$ in $\boldsymbol{P}(S)$. This section explores connections between this convergence concept and those of Section 1.5.

Theorem 2.2.1 *Suppose that the $\{X_n\}$ above are defined on a common probability space and $X_n \to X_\infty$ in probability. Then $X_n \to X_\infty$ in law.*

Proof Let $f : S \to \boldsymbol{R}$ be bounded and uniformly continuous with respect to d. Let $\epsilon > 0$ and pick $\delta > 0$ such that $d(x, y) < \delta$ implies $| f(x) - f(y) | < \epsilon$ for $x, y \in S$. Then

$$\begin{aligned}
\left| \int f d\mu_n - \int f d\mu_\infty \right| &= | E[f(X_n)] - E[f(X_\infty)] | \\
&\leq E[| f(X_n) - f(X_\infty) | I\{d(X_n, X_\infty) < \delta\}] \\
&\quad + E[| f(X_n) - f(X_\infty) | I\{d(X_n, X_\infty) \geq \delta\}] \\
&\leq \epsilon + 2 \| f \| P(d(X_n, X_\infty) \geq \delta) \\
&\to \epsilon \text{ as } n \to \infty.
\end{aligned}$$

Since $\epsilon > 0$ was arbitrary, $\int f d\mu_n \to \int f d\mu_\infty$. Use Theorem 2.1.1(iii) to conclude. □

A converse to this statement does not in general make sense as $\{X_n\}$ need not even be defined on the same probability space. A sort of converse is provided by the following important result of Skorohod.

Theorem 2.2.2 *Let $\mu_n \to \mu_\infty$ in $\boldsymbol{P}(S)$. Then there exists a probability space on which there are S-valued random variables $X_n, n = 1, 2, \cdots, \infty$, such that $\mathcal{L}(X_n) = \mu_n$ for all n and $X_n \to X_\infty$ a.s.*

Proof Take $\Omega = [0,1), \mathcal{F}$ its Borel σ-field and P the Lebesgue measure. To every ordered finite collection $\{i_1, i_2, \cdots, i_k\}, k \geq 1$, of natural numbers, associate a Borel set $S_{(i_1,\cdots,i_k)}$ in S as follows:

(i) If $(i_1, \cdots, i_k) \neq (j_1, \cdots, j_k)$, then $S_{(i_1,\cdots,i_k)} \bigcap S_{(j_1,\cdots,j_k)} = \phi$,

(ii) $\bigcup_j S_j = S$, $\bigcup_j S_{(i_1,\cdots,i_k,j)} = S_{(i_1,\cdots,i_k)}$,

(iii) Diameter $(S_{(i_1,\cdots,i_k)}) < 2^{-k}$ under the metric d,

(iv) $\mu_n(\partial S_{(i_1,\cdots,i_k)}) = 0$ for $n = 1, 2, \cdots, \infty$.

Thus for each k, $\{S_{(i_1,\cdots,i_k)}\}$ is a disjoint cover of S which is a refinement of the corresponding cover for $k' < k$. One way to obtain these sets is as follows: For each k, let $B_{mk}, m = 1, 2, \cdots$, be open balls of radius not exceeding $2^{-(k+1)}$, covering S and satisfying $\mu_n(\partial B_{mk}) = 0$ for all n, k, m. This is always possible (Exercise 2.5). Let $D_{1k} = B_{1k}, D_{nk} = B_{nk} \backslash (\bigcup_{m=1}^{n-1} B_{mk}), S_{(i_1,\cdots,i_k)} = \bigcap_{j=1}^{k} D_{i_j j}$. Then (i)–(iv) hold. For each k, order $\{(i_1, \cdots, i_k)\}$ lexicographically. Define intervals $\Delta^n(i_1, \cdots, i_k)$ of the form $[a, b)$ in $[0, 1)$ such that

(a) The length of $\Delta^n(i_1, \cdots, i_k)$ is $\mu_n(S_{(i_1,\cdots,i_k)})$ for $n = 1, 2, \ldots, \infty$,

(b) $\bigcup_{(i_1,\cdots,i_k)} \Delta^n(i_1, \cdots, i_k) = [0, 1)$ for $n = 1, 2, \cdots, \infty$ and

(c) when $(i_1, \cdots, ik) < (j_1, \cdots, j_k)$, the left end point of $\Delta^n(j_1, \cdots, j_k)$ lies to the right of the interval $\Delta^n(i_1, \cdots, i_k)$ for $n = 1, 2, \cdots, \infty$.

These intervals are uniquely determined. For each $S_{(i_1,\cdots,i_k)}$ such that $\mu_n(S_{(i_1,\cdots,i_k)}) > 0$ for some n, the interior is nonempty by (iv) above. Thus we may take a point $x_{(i_1,\cdots,i_k)}$ in its interior. For $\omega \in \Omega$, define

$$X_n^k(\omega) = x_{(i_1,\cdots,i_k)} \text{ on } \Delta^n(i_1, \cdots, i_k)(\omega),\ k \geq 1,\ n = 1, 2, \cdots \infty.$$

Then $d(X_n^k(\omega), X_n^{k+p}(\omega)) \leq 2^{-k}, k, p \geq 1, n = 1, 2, \cdots, \infty$, making $\{X_n^k(\omega)\}$ Cauchy for each n, ω. Thus $X_n(\omega) = \lim_{k\to\infty} X_n^k(\omega)$ exists for all n, ω. Letting $| I |$ denote the length of the interval I, we have

$$\mu_n(S_{(i_1,\cdots,i_k)}) = | \Delta^n(i_1, \cdots, i_k) | \rightarrow | \Delta^\infty(i_1, \cdots, i_k) | = \mu_\infty(S_{(i_1,\cdots,i_k)})$$

by Theorem 2.1.1(vi) and (iv) above. Hence for each ω in the interior of $\Delta^\infty(i_1, \cdots, i_k)$, there exists an integer $n_k(\omega) \geq 1$ such that ω lies in the interior of $\Delta^n(i_1, \cdots, i_k)$ for $n \geq n_k(\omega)$. Then $X_n^k(\omega) = X_\infty^k(\omega)$ for $n \geq n_k(\omega)$, implying

$$\begin{aligned} & d(X_n(\omega), X_\infty(\omega)) \\ \leq \; & d(X_n(\omega), X_n^k(\omega)) + d(X_n^k(\omega), X_\infty^k(\omega)) + d(X_\infty^k(\omega), X_\infty(\omega)) \\ \leq \; & 2^{-k+1}, \; n \geq n_k(\omega). \end{aligned}$$

Setting

$$\Omega_o = \bigcap_{k=1}^{\infty} \bigcup_{(i_1,\cdots,i_k)} \text{interior}\Delta^\infty(i_1, \cdots, i_k),$$

$P(\Omega_o) = 1$ and $X_n(\omega) \rightarrow X_\infty(\omega)$ for $\omega \in \Omega_o$. Now for each $p \geq 1$,

$$P(X_n^{k+p}(\omega) \in \overline{S}_{(i_1,\cdots,i_k)}) = P(X_n^{k+p}(\omega) \in \text{int}(S_{(i_1,\cdots,i_k)})) = \mu_n(S_{(i_1,\cdots,i_k)}).$$

Also, every open set in S is expressible as a disjoint countable union of some $S_{(i_1,\cdots,i_k)}$'s. It follows (Exercise 2.6) that for every open $G \subset S$,

$$\liminf_{k\to\infty} P(X_n^k \in G) \geq \mu_n(G).$$

By Theorem 2.1.1(v), $\mathcal{L}(X_n^k) \rightarrow \mu_n$. But $X_n^k \rightarrow X_n$ a.s. Thus $\mathcal{L}(X_n) = \mu_n$ for all $n = 1, 2, \cdots, \infty$. This completes the proof. □

This result allows us to deduce several consequences of convergence in law from known properties of a.s. convergence. The following corollary gives an instance of this.

Corollary 2.2.1 *For $\{X_n\}$ as above with $S = \boldsymbol{R}$, let $X_n \rightarrow X_\infty$ in law. If $\{X_n\}$ are u.i., then $E[X_n] \rightarrow E[X_\infty]$.*

The proof is easy (Exercise 2.7).

See [5, 47] for some refinements of Skorohod's theorem.

2.3 Compactness in $\mathbb{P}(S)$

The following theorem of Prohorov gives a useful characterization of relative compactness in $\boldsymbol{P}(S)$.

Theorem 2.3.1 *$\Lambda \subset \boldsymbol{P}(S)$ is relatively compact if and only if it is tight.*

Proof Suppose S is compact. By the Riesz theorem, $\boldsymbol{P}(S) = \{\mu \in C^*(S) \mid \int f d\mu \geq 0$ for $f \geq 0$ and $\mu(S) = 1\}$ with the weak*-topology of $C^*(S)$ relativized to it. It is clearly bounded and weak*-closed in $C^*(S)$ and therefore compact by the Banach–Alaoglu theorem.

Suppose S is not compact. Recall the map $h : S \to [0,1]^\infty$ of Theorem 1.1.1. Let $\Lambda \subset \boldsymbol{P}(S)$ be tight and $\{\mu_n, n \geq 1\} \subset \Lambda$. For each $n \geq 1$, define $\overline{\mu}_n \in \boldsymbol{P}(\overline{h(S)})$ by $\overline{\mu}_n(A) = \mu_n(h^{-1}(A))$ for A Borel in $\overline{h(S)}$. By the foregoing, $\{\overline{\mu}_n\}$ is relatively compact in $\boldsymbol{P}(\overline{h(S)})$ and therefore for any subsequence of $\{n\}$, denoted $\{n\}$ again by abuse of notation, there exists a further subsequence $\{n(k)\}$ such that $\overline{\mu}_{n(k)} \to \overline{\mu}$ in $\boldsymbol{P}(\overline{h(S)})$ for some $\overline{\mu}$. For $m \geq 1$, let $K_m \subset S$ be compact such that $\mu_n(K_m) > 1 - \frac{1}{m}$ for all n. Then $h(K_m), m \geq 1$, are compact and for each m,

$$\begin{aligned}\overline{\mu}(h(K_m)) &\geq \limsup_{k\to\infty} \overline{\mu}_{n(k)}(h(K_m)) \\ &= \limsup_{k\to\infty} \mu_{n(k)}(K_m) \geq 1 - \frac{1}{m}\,.\end{aligned}$$

Thus $\overline{\mu}(\bigcup_m h(K_m)) = 1 = \overline{\mu}(h(S))$ and $\overline{\mu}$ restricts to some $\mu' \in \boldsymbol{P}(h(S))$ on $h(S)$. Let μ be the image of μ' under h^{-1}. Then

$$\limsup_{k\to\infty} \mu_{n(k)}(F) = \limsup_{k\to\infty} \overline{\mu}_{n(k)}(h(F)) \leq \overline{\mu}(h(F)) = \mu(F)$$

for all closed $F \subset S$, implying $\mu_{n(k)} \to \mu$ in $\boldsymbol{P}(S)$. Thus Λ is relatively sequentially compact and hence relatively compact.

Conversely, let Λ be relatively compact. Let $\{s_i, i \geq 1\} \subset S$ be countable dense. For $k, n \geq 1$, define $G_k^n = \bigcup_{j=1}^n B(s_j, 1/k)$ where $B(x,r)$ is the open ball of radius r and centre x. The maps $\mu \to \mu(G_k^n)$ are lower semicontinuous by Theorem 2.1.1(v) and increase to the constant function 1 as $n \to \infty$. By Dini's theorem, this convergence is uniform on the compact set $\overline{\Lambda}$. Thus for $\epsilon > 0, k \geq 1$, there is an $n(k) \geq 1$ such that

$$\inf_{\mu\in\overline{\Lambda}} \mu(G_k^{n(k)}) \geq 1 - \epsilon/2^k.$$

Then $K = \overline{\bigcap_{k=1}^\infty G_k^{n(k)}}$ is totally bounded and closed. Therefore it is compact. It clearly satisfies $\inf_{\mu\in\overline{\Lambda}} \mu(K) \geq 1 - \epsilon$. Thus Λ is tight. □

The next result gives a useful criterion for relative compactness in $\boldsymbol{P}(S)$.

Theorem 2.3.2 *Let $\{\mu_\alpha, \alpha \in I\} \in \boldsymbol{P}(S)$ satisfy: $\mu_\alpha \ll \mu$ for all $\alpha \in I$ and a prescribed $\mu \in \boldsymbol{P}(S)$. Suppose $\frac{d\mu_\alpha}{d\mu}, \alpha \in I$, are u.i. with respect to μ. Then $\{\mu_\alpha, \alpha \in I\}$ is relatively compact.*

This is immediate from Theorem 1.3.5. Another related result is the following theorem due to Scheffe.

Theorem 2.3.3 *Let $\mu_n, n = 1, 2, \cdots, \infty$ be probability measures on a measurable space $(\Omega, \mathcal{F})$ such that $\mu_n \ll \lambda, n = 1, 2, \cdots, \infty$, for some nonnegative $\sigma-$finite measure λ on $(\Omega, \mathcal{F})$. If*

$$\frac{d\mu_n}{d\lambda} \to \frac{d\mu_\infty}{d\lambda}, \ \lambda - a.s.,$$

then $\mu_n \to \mu_\infty$ in total variation.

Proof The claim amounts to showing that

$$\frac{d\mu_n}{d\lambda} \to \frac{d\mu_\infty}{d\lambda} \text{ in } L_1(\Omega, \mathcal{F}, \lambda).$$

This follows as in Theorem 1.3.3. □

For S countable with the discrete topology, this shows that convergence in $\boldsymbol{P}(S)$ and in total variation norm are equivalent.

2.4 Complete Metrics on $\mathbb{P}(S)$

The metric ρ above is often convenient to work with, but need not be complete as the following result shows.

Theorem 2.4.1 *ρ is complete if and only if S is compact.*

Proof Let φ be the map that maps $\mu \in \boldsymbol{P}(S)$ to $[\int f_1 d\mu, \int f_2 d\mu, \cdots] \in [-1, 1]^\infty$, where $\{f_i\}$ are as in the definition of ρ. Argue as in the proof of Theorem 1.1.1 to conclude that φ is a homeomorphism onto a subset of $[-1, 1]^\infty$. Metrize $[-1, 1]^\infty$ by the metric

$$\overline{\rho}((x_1, x_2, \cdots), (y_1, y_2, \cdots)) = \sum_{n=1}^{\infty} 2^{-n} \mid x_n - y_n \mid .$$

Then φ is an isometry between $(\boldsymbol{P}(S), \rho)$ and $(\varphi(\boldsymbol{P}(S)), \overline{\rho})$. Thus ρ is complete if and only if $\overline{\rho}$ is a complete metric on $\varphi(\boldsymbol{P}(S))$, which is true if and only if $\varphi(\boldsymbol{P}(S))$ is closed and therefore compact in $[-1, 1]^\infty$. Since φ is a homeomorphism, the latter holds if and only if $\boldsymbol{P}(S)$ is compact. By Prohorov's theorem, $\boldsymbol{P}(S)$ is compact if S is.

Conversely, if S is not compact, it contains a sequence $\{s_n\}$ that has no limit point. Then the Dirac measures at $s_n, n \geq 1$, have no limit point either, proving that $\boldsymbol{P}(S)$ is not compact. This completes the proof. □

We shall describe below two complete metrics on $\boldsymbol{P}(S)$, implying in particular that $\boldsymbol{P}(S)$ is Polish. The first is the Prohorov metric q defined as follows: For $\epsilon > 0$ and $A \subset S$, let $A^\epsilon = \{x \in S \mid d(x, A) < \epsilon\}$. For $\mu, \nu \in \boldsymbol{P}(S)$, define

$$q(\mu, \nu) = \inf\{\epsilon > 0 \mid \mu(A) \leq \nu(A^\epsilon) + \epsilon, \nu(A) \leq \mu(A^\epsilon) + \epsilon \text{ for Borel } A \subset S\}.$$

An alternative definition is given by

$$q(\mu, \nu) = \inf\{\epsilon > 0 \mid \mu(F) \leq \nu(F^\epsilon) + \epsilon \text{ for all closed } F \subset S\}.$$

Lemma 2.4.1 *The two definitions above are equivalent.*

Proof Let $\epsilon > 0$ be such that $\mu(F) \leq \nu(F^\epsilon) + \epsilon$ for all closed $F \subset S$. We shall first prove that $\nu(F) \leq \mu(F^\epsilon) + \epsilon$ for all closed $F \subset S$. Let $H \subset S$ be closed and $G = S \backslash H^\epsilon$. Then $H \subset S \backslash G^\epsilon$ and hence

$$\mu(H^\epsilon) = 1 - \mu(G) \geq 1 - \nu(G^\epsilon) - \epsilon \geq \nu(H) - \epsilon,$$

proving the claim. Thus it suffices to prove that if $\epsilon > 0$ satisfies

$$\mu(A) \leq \nu(A^\epsilon) + \epsilon, \ \nu(A) \leq \mu(A^\epsilon) + \epsilon,$$

for all Borel A, it does so for all closed A and vice versa. The first implication is trivial. Conversely, let the above inequalities hold for all closed A. Let $B \subset S$ be Borel. Pick $\delta > 0$. By Exercise 1.20, there exists a closed set $C \subset B$ such that $\mu(C) \geq \mu(B) - \delta$. Thus

$$\mu(B) - \delta \leq \mu(C) \leq \nu(C^\epsilon) + \epsilon \leq \nu(B^\epsilon) + \epsilon.$$

Since $\delta > 0$ was arbitrary, $\mu(B) \leq \nu(B^\epsilon) + \epsilon$. The other inequality follows similarly. □

Theorem 2.4.2 *q defines a metric on $\boldsymbol{P}(S)$ consistent with the Prohorov topology.*

Proof Clearly, $q(\mu, \nu) = q(\nu, \mu) \geq 0 = q(\mu, \mu)$. Also, for closed $A \subset S, \mu(A^\epsilon) \downarrow \mu(A)$ as $\epsilon \downarrow 0$. Thus $q(\mu, \nu) = 0$ implies $\mu(A) = \nu(A)$ for closed A and therefore $\mu = \nu$. Finally, let $\mu_1, \mu_2, \mu_3 \in \boldsymbol{P}(S)$. For any $\epsilon > q(\mu_1, \mu_2)$ and $\delta > q(\mu_2, \mu_3)$, we have

$$\mu_1(A) < \mu_2(A^\epsilon) + \epsilon, \ \mu_2(A^\epsilon) < \mu_3(A^{\epsilon+\delta}) + \delta$$

for Borel $A \subset S$, implying $\mu_1(A) < \mu_3(A^{\epsilon+\delta}) + \epsilon + \delta$. Similarly, $\mu_3(A) \leq \mu_1(A^{\epsilon+\delta}) + \epsilon + \delta$. Thus

$$q(\mu_1, \mu_3) \leq \epsilon + \delta.$$

Letting $\epsilon \to q(\mu_1, \mu_2), \delta \to q(\mu_2, \mu_3)$, we get

$$q(\mu_1, \mu_3) \leq q(\mu_1, \mu_2) + q(\mu_2, \mu_3).$$

Thus q is a metric.

Let $\mu \in \boldsymbol{P}(S), F \subset S$ closed and $\epsilon > 0$. Let $\delta \in (0, \epsilon)$ be such that $\mu(F^\delta) < \mu(F) + \epsilon$. If $q(\mu, \nu) < \delta$, then $\nu(F) < \mu(F^\delta) + \delta < \mu(F) + 2\epsilon$. Thus each set of type (iii) in Corollary 2.1.1 contains a $q-$ball.

Now pick $\delta \in (0, \epsilon/2)$. Cover S by open balls $\{S_i\}$ satisfying $\mu(\partial S_i) = 0$ and diameter$(S_i) < \delta$ for all i. Let $A_1 = S_1, A_n = S_n \backslash (\bigcup_{m=1}^{n-1} S_m), n \geq 2$. Take $k \geq 1$ such that $\mu(\bigcup_{i=1}^k A_i) > 1 - \delta$. Since $\mu(\partial S_i) = 0$ for all i, $\mu(\partial(\bigcup_{i=1}^k A_i)) = 0$. Let N be a neighborhood of μ of type (v) in Corollary 2.1.1 described by $N = \{\nu \in \boldsymbol{P}(S) | \mid \mu(A) - \nu(A) \mid < \delta$ for all A that can be written as a union of sets from $\{A_1, \cdots, A_k\}\}$. Then $\nu(\bigcup_{i=1}^k A_i) > 1 - 2\delta$ for $\nu \in N$. For any Borel set $B \subset S$, let $A' =$ the union of sets in $\{A_1, \cdots, A_k\}$ which intersect B. Then $\mid \mu(A') - \nu(A') \mid < \delta$ for $\nu \in N$. But $B \subset A' \bigcup (\bigcup_{i=1}^k A_i)^c$. Since diameter $(A_i) < \delta$ for all $i, A' \subset B^\delta$ and thus

$$\mu(B) \leq \mu(A') + \mu((\bigcup_{i=1}^k A_i)^c) \leq \nu(A') + 2\delta \leq \nu(B^\delta) + 2\delta.$$

Thus $q(\mu, \nu) \leq 2\delta < \epsilon$. That is, every open q-ball contains a set of type (v) of Corollary 2.1.1. This completes the proof. □

Theorem 2.4.3 *q is a complete metric on $\boldsymbol{P}(S)$.*

Proof Let $\{\mu_n\}$ be a Cauchy sequence in $\boldsymbol{P}(S)$ with respect to q. Let $\epsilon > 0$ and $k \geq 1$. Pick $n(k) \geq 1$ such that for $n \geq n(k), q(\mu_n, \mu_{n(k)}) < \epsilon 2^{-k}$. Let $B_1^k, \cdots, B_{m(k)}^k$ be finitely many open balls of diameter $\epsilon 2^{-k}$ satisfying

$$\mu_{n(k)} \left(\bigcup_{i=1}^{m(k)} B_i^k \right) > 1 - \epsilon 2^{-k}.$$

Let $A_i^k, 1 \leq i \leq m(k)$, be open balls concentric with the respective B_i^k's and with twice the radius. Since

$$\left(\bigcup_{i=1}^{m(k)} B_i^k \right)^{\epsilon 2^{-k}} \subset \bigcup_{i=1}^{m(k)} A_i^k$$

and $q(\mu_n, \mu_{n(k)}) \leq \epsilon 2^{-k}$ for $n \geq n(k)$, we have

$$\mu_n \left(\bigcup_{i=1}^{m(k)} A_i^k \right) > 1 - \epsilon 2^{-k+1}, n \geq n(k).$$

By including finitely many additional open balls of radius $\epsilon 2^{-k+1}$ if necessary, denoted by, say, $A^k_{m(k)+1}, \cdots, A^k_{j(k)}$, we have

$$\mu_n(\bigcup_{i=1}^{j(k)} A^k_i) > 1 - \epsilon 2^{-k+1}, n \geq 1.$$

Let $K = \bigcap_{k\geq 1} \bigcup_{i \leq j(k)} A^k_i$. Then $\overline{K}$ is totally bounded and closed, hence compact. Also, $\mu_n(\overline{K}) \geq 1 - 2\epsilon$ for all n. Thus $\{\mu_n\}$ is tight and therefore relatively compact. Since it is Cauchy as well, it must converge. □

The metric q has the probabilistic interpretation (due to Strassen)

$$q(\mu, \nu) = \inf_M [\inf\{\epsilon > 0 \mid P(d(X, Y) \geq \epsilon) \leq \epsilon\}],$$

where M = the set of all of $S \times S$-valued random variables (X, Y) satisfying $\mathcal{L}(X) = \mu, \mathcal{L}(Y) = \nu$. We shall not prove this, referring the reader to [19, pp.96-101], for details. Instead we introduce another related metric, appealing because of its probabilistic interpretation. This is

$$\overline{q}(\mu, \nu) = \inf_M E[d(X, Y)] \tag{2.1}$$

for M as above.

Theorem 2.4.4 *$\overline{q}$ is a complete metric on $\boldsymbol{P}(S)$ equivalent to q.*

To prove this, we need the following lemma, the proof of which borrows a result from the next chapter.

Lemma 2.4.2 *Let $\mu_1, \mu_2 \in \boldsymbol{P}(S \times S)$ be such that $\mu_1(S, dy), \mu_2(dy, S) \in \boldsymbol{P}(S)$ coincide ($= \nu$, say). Then there exist S-valued random variables (X, Y, Z) on some probability space $(\Omega, \mathcal{F}, P)$ such that $\mathcal{L}((X, Y)) = \mu_1$ and $\mathcal{L}((Y, Z)) = \mu_2$.*

Proof Anticipating a result in the next chapter (Corollary 3.1.2), we "disintegrate" μ_1, μ_2 as: $\mu_1(dx, dy) = \nu(dy)v_1(y, dx), \mu_2(dy, dz) = \nu(dy)v_2(y, dz)$ for ν as above and $y \to v_1(y, .), y \to v_2(y, .) : S \to \boldsymbol{P}(S)$ measurable. Let $\Omega = S \times S \times S$ with $\mathcal{F}$ its Borel σ-field and P a probability measure on $(\Omega, \mathcal{F})$ defined by

$$P(A \times B \times C) = \int_B \int_A \int_C v_2(y, dz) v_1(y, dx) \nu(dy)$$

for A, B, C Borel in S. Letting $\omega = (\omega_1, \omega_2, \omega_3)$ denote a typical element of Ω, the random variables X, Y, Z defined by $X(\omega) = \omega_1, Y(\omega) = \omega_2, Z(\omega) = \omega_3$ will do. □

Proof of Theorem 2.4.4 Clearly, $\overline{q}(\mu,\nu) = \overline{q}(\nu,\mu) \geq 0 = \overline{q}(\mu,\mu)$ (Take $X = Y$ in (2.1).). For $\mu, \nu \in \boldsymbol{P}(S)$, let X, Y be random variables on some probability space with $\mathcal{L}(X) = \mu, \mathcal{L}(Y) = \nu$. Then for Borel $A \subset S$ and $\delta > 0$,

$$\begin{aligned} \mu(A) - \nu(A^\delta) &= E[I\{X \in A\} - I\{Y \in A^\delta\}] \\ &\leq P(X \in A, Y \notin A^\delta) \\ &\leq P(d(X,Y) \geq \delta) \\ &\leq \delta^{-1} E[d(X,Y)]. \end{aligned}$$

Thus

$$\mu(A) - \nu(A^\delta) \leq \delta^{-1}\overline{q}(\mu,\nu). \tag{2.2}$$

Since $\delta > 0$ is arbitrary, $\overline{q}(\mu,\nu) = 0$ implies $q(\mu,\nu) = 0$ and hence $\mu = \nu$. Now let $\mu_1, \mu_2, \mu_3 \in \boldsymbol{P}(S)$ and $(X', Y'), (Y'', Z'')$ $S \times S$-valued random variables (possibly on different probability spaces) such that $\mathcal{L}(X') = \mu_1, \mathcal{L}(Y') = \mathcal{L}(Y'') = \mu_2, \mathcal{L}(Z'') = \mu_3$ and

$$\begin{aligned} E[d(X',Y')] &\leq \overline{q}(\mu_1,\mu_2) + \epsilon, \\ E[d(Y'',Z'')] &\leq \overline{q}(\mu_2,\mu_3) + \epsilon, \end{aligned}$$

for some prescribed $\epsilon > 0$. By the preceding lemma, we can construct on a common probability space random variables X, Y, Z such that $\mathcal{L}((X,Y)) = \mathcal{L}((X',Y'))$ and $\mathcal{L}((Y,Z)) = \mathcal{L}((Y'',Z''))$. Then

$$\begin{aligned} \overline{q}(\mu_1,\mu_3) &\leq E[d(X,Z)] \\ &\leq E[d(X,Y)] + E[d(Y,Z)] \\ &\leq \overline{q}(\mu_1,\mu_2) + \overline{q}(\mu_2,\mu_3) + 2\epsilon. \end{aligned}$$

Letting $\epsilon \to 0$, we get the triangle inequality. Thus $\overline{q}$ is a metric.

If $\overline{q}(\mu_n, \mu_\infty) \to 0$, setting $\mu = \mu_n$ and $\nu = \mu_\infty$ in (2.2), we conclude that the right hand side of (2.2) can be made less than δ for sufficiently large n, implying that $q(\mu_n, \mu_\infty) \leq \delta$ for such n. Since $\delta > 0$ was arbitrary, this implies that $q(\mu_n, \mu_\infty) \to 0$. A similar argument shows that if $\{\nu_n\} \subset \boldsymbol{P}(S)$ is Cauchy with respect to $\overline{q}$, it is so with respect to q and hence converges. Finally, let $q(\eta_n, \eta_\infty) \to 0$. Then by Skorohod's theorem, there exist on some probability space S-valued random variables $X_n, n = 1, 2, \cdots, \infty$, such that $\mathcal{L}(X_n) = \eta_n$ for $n = 1, 2, \cdots, \infty$ and $X_n \to X_\infty$ a.s. Thus

$$E[d(X_n, X_\infty)] \to 0,$$

implying $\overline{q}(\eta_n, \eta_\infty) \to 0$. This completes the proof. □

It should be remarked parenthetically that the fact that the complete metric d is bounded plays a role in the definition of $\overline{q}$ above, but not in the definition of q.

2.5 Characteristic Functions

Characteristic functions provide a useful tool for ascertaining convergence in $\boldsymbol{P}(\boldsymbol{R}^m)$ for $m \geq 1$. The characteristic function of $\mu \in \boldsymbol{P}(\boldsymbol{R}^m)$ is simply its Fourier transform $\varphi_\mu : \boldsymbol{R}^m \to \boldsymbol{C}$ given by $\varphi_\mu(t) = \int exp(i\langle t, z\rangle)\mu(dz)$. The following properties are immediate from the definition (Exercise 2.8). Let $\mathcal{L}(X) = \mu$.

(i) $\varphi_\mu(0) = 1, | \varphi_\mu(t) | \leq 1, \varphi_\mu(t) = \overline{\varphi_\mu(-t)}$.

(ii)

$$\begin{aligned} | \varphi_\mu(t+h) - \varphi_\mu(t) | &\leq E[| exp(i\langle t+h, X\rangle) - exp(i\langle t, X\rangle) |] \\ &= E[| exp(i\langle h, X\rangle) - 1 |] \to 0 \end{aligned}$$

as $\| h \| \to 0$. Thus φ_μ is uniformly continuous.

(iii) For $a \in \boldsymbol{R}, b \in \boldsymbol{R}^m, \nu = \mathcal{L}(aX+b), \varphi_\nu(t) = exp(i\langle t, b\rangle)\varphi_\mu(at)$.

(iv) φ_μ is positive definite, i.e., for $n \geq 1, x_1, .., x_n \in \boldsymbol{R}^m$ and $c_1, \cdots, c_n \in \boldsymbol{C}$,

$$\sum_{j,k=1}^{n} c_j \varphi_\mu(x_j - x_k)\bar{c}_k \geq 0.$$

The importance of characteristic functions in probability theory stems from the Lévy continuity theorem which is Theorem 2.5.1 below. We start with some preliminaries.

Lemma 2.5.1 *Let $\mu, \nu \in \boldsymbol{P}(\boldsymbol{R}^m)$ be such that $\varphi_\mu = \varphi_\nu$. Then $\mu = \nu$.*

Proof We shall use the identity

$$\frac{1}{(2\pi)^m} \int e^{i\langle x,y\rangle - \frac{1}{2}\sigma^2\|y\|^2} dy = \frac{1}{\sqrt{(2\pi)^m \sigma^{2m}}} e^{-\|x\|^2/2\sigma^2}, \sigma > 0.$$

Replace x by $x - z$, integrate both sides with respect to $\mu(dx)$ and use Fubini's theorem to deduce

$$\frac{1}{(2\pi)^m} \int \varphi_\mu(y) e^{-i\langle z,y\rangle} e^{-\frac{1}{2}\sigma^2\|y\|^2} dy = \frac{1}{\sqrt{(2\pi)^m \sigma^{2m}}} \int e^{-\|x-z\|^2/2\sigma^2} \mu(dx) .$$

A similar identity holds for ν. Since $\varphi_\mu = \varphi_\nu$, we have

$$\frac{1}{\sqrt{(2\pi)^m \sigma^{2m}}} \int e^{-\|x-z\|^2/2\sigma^2} \mu(dx) = \frac{1}{\sqrt{(2\pi)^m \sigma^{2m}}} \int e^{-\|x-z\|^2/2\sigma^2} \nu(dx) .$$

Multiply both sides by $f(z), f \in C_b(\boldsymbol{R}^m)$, and integrate over $\boldsymbol{R}^m$ to obtain (after a change of variables)

$$\frac{1}{\sqrt{(2\pi)^m\sigma^{2m}}}\int\int f(x+y)e^{-\|y\|^2/2\sigma^2}\mu(dx)dy$$

$$= \frac{1}{\sqrt{(2\pi)^m\sigma^{2m}}}\int\int f(x+y)e^{-\|y\|^2/2\sigma^2}\nu(dx)dy .$$

Use Fubini's theorem to interchange orders of integration and let $\sigma^2 \to 0$ to obtain (Exercise 2.9) $\int f d\mu = \int f d\nu$. Thus $\mu = \nu$. □

Corollary 2.5.1 *If X, Y are independent $\boldsymbol{R}^m$-valued random variables with $\mathcal{L}(X) = \mu, \mathcal{L}(Y) = \nu, \mathcal{L}(X+Y) = \xi$, then $\xi = \mu * \nu$, i.e., the convolution of μ and ν, defined by*

$$\mu * \nu(B) = \int \mu(B-x)\nu(dx) = \int \nu(B-x)\mu(dx) \textit{ for } B \textit{ Borel in } \boldsymbol{R}^m.$$

Proof By independence of $X, Y, \varphi_\xi = \varphi_\mu\varphi_\nu$. It is easily verified that $\varphi_{\mu*\nu} = \varphi_\mu\varphi_\nu$. The claim follows. □

Theorem 2.5.1 *(Lévy continuity theorem) If $\mu_n \to \mu_\infty$ in $\boldsymbol{P}(\boldsymbol{R}^m)$, then $\varphi_{\mu_n} \to \varphi_{\mu_\infty}$ uniformly on compacts. Conversely, if $\varphi_{\mu_n} \to \varphi$ pointwise for some $\varphi : \boldsymbol{R}^m \to \boldsymbol{C}$ which is continuous at zero, then $\varphi = \varphi_{\mu_\infty}$ for some $\mu_\infty \in \boldsymbol{P}(\boldsymbol{R}^m)$ and $\mu_n \to \mu_\infty$ in $\boldsymbol{P}(\boldsymbol{R}^m)$.*

Proof Let $\mu_n \to \mu_\infty$. By Theorem 2.2.2, there exist $\boldsymbol{R}^m$-valued random variables $X_n, n = 1, 2, \cdots, \infty$, on some probability space such that $X_n \to X_\infty$ a.s. and $\mathcal{L}(X_n) = \mu_n$ for each n. For compact $K \subset \boldsymbol{R}^n$ it is easily seen that

$$\lim_{n\to\infty} \sup_{t\in K} | exp(i\langle t, X_n\rangle) - exp(i\langle t, X_\infty\rangle) | = 0, \text{ a.s.},$$

from which the first claim follows.

For the converse, let $\overline{\boldsymbol{R}}^m = \boldsymbol{R}^m \bigcup\{\infty\}$ be the one point compactification of $\boldsymbol{R}^m$. Extend $\mu_n \in \boldsymbol{P}(\boldsymbol{R}^m)$ to $\overline{\mu}_n \in \boldsymbol{P}(\overline{\boldsymbol{R}}^m)$ by setting $\overline{\mu}_n(\{\infty\}) = 0, n \geq 1$. Then $\{\overline{\mu}_n\}$ are relatively compact in $\boldsymbol{P}(\overline{\boldsymbol{R}}^m)$ and converge along a subsequence $\{n(k)\}$ to some $\overline{\mu} \in \boldsymbol{P}(\overline{\boldsymbol{R}}^m)$. For $\delta > 0$, let $A_\delta = \{x = (x_1, \cdots, x_m) \in \boldsymbol{R}^m \mid \; | x_i | \leq \delta, 1 \leq i \leq m\}$. Then a simple computation yields

$$\begin{aligned}\frac{1}{(2\delta)^m}\int_{A_\delta}\varphi_{\mu_{n(k)}}(t)dt &= \int_{R^m}\prod_{i=1}^m \frac{sin\delta x_i}{\delta x_i}d\mu_{n(k)}(x)\\ &= \int_{\overline{R}^m} f_\delta(x)d\overline{\mu}_{n(k)}(x)\end{aligned}$$

where $x = (x_1, \cdots, x_m)$ and $f_\delta(x)$ is the continuous extension of $x \to \Pi_i(sin\delta x_i/\delta x_i)$ to $\overline{\boldsymbol{R}}^m$. Clearly, $f_\delta(\infty) = 0$. Now

$$1 = \varphi(0) = \lim_{\delta \to 0} \frac{1}{(2\delta)^m} \int_{A_\delta} \varphi(t)dt.$$

By the dominated convergence theorem,

$$\begin{aligned}
\frac{1}{(2\delta)^m} \int_{A_\delta} \varphi(t)dt &= \lim_{k\to\infty} \frac{1}{(2\delta)^m} \int_{A_\delta} \varphi_{n(k)}(t)dt \\
&= \lim_{k\to\infty} \frac{1}{(2\delta)^m} \int_{R^m} \prod_{i=1}^{m} \frac{sin\delta x_i}{\delta x_i} d\mu_{n(k)}(t) \\
&= \lim_{k\to\infty} \int_{\overline{R}^m} f_\delta(x) d\overline{\mu}_{n(k)}(x) \\
&= \int_{\overline{R}^m} f_\delta(x) d\overline{\mu}(x) \\
&= \int_{R^m} f_\delta(x) d\mu(x)
\end{aligned}$$

where μ is the restriction of $\overline{\mu}$ to $\boldsymbol{R}^m$. Letting $\delta \to 0$ and using the fact that $f_\delta(x) \to 1$ as $\delta \to 0$ for $x \in \boldsymbol{R}^m$ in a uniformly bounded manner, we have

$$1 = \int_{R^m} d\mu(x),$$

that is, $\mu \in \boldsymbol{P}(\boldsymbol{R}^m)$. Since $\overline{\mu}_{n(k)} \to \overline{\mu}$ in $\boldsymbol{P}(\overline{\boldsymbol{R}}^m)$,

$$\begin{aligned}
\text{liminf}_{k\to\infty}\mu_{n(k)}(G) &= \text{liminf}_{k\to\infty}\overline{\mu}_{n(k)}(G) \\
&\geq \overline{\mu}(G) = \mu(G)
\end{aligned}$$

for all open $G \subset \boldsymbol{R}^m$, implying $\mu_{n(k)} \to \mu$ in $\boldsymbol{P}(\boldsymbol{R}^m)$. Hence $\varphi = \varphi_\mu$. This argument shows that $\{\mu_n\}$ is relatively compact and every limit point μ thereof satisfies $\varphi_\mu = \varphi$. The claim now follows from Lemma 2.5.1. □

The next result distinguishes characteristic functions.

Theorem 2.5.2 *(Bochner) A function $\varphi : \boldsymbol{R}^m \to \boldsymbol{C}$ is a characteristic function of some $\mu \in \boldsymbol{P}(\boldsymbol{R}^m)$ if and only if it satisfies the properties (i), (ii) and (iv) above.*

Proof Only the "if" part needs to be proved. Let φ satisfy (i), (ii), (iv). Assume in addition that $|\varphi| \in L_1(\boldsymbol{R}^m)$. Let $g : \boldsymbol{R}^m \to \boldsymbol{C}$ be bounded continuous with $|g| \in L_1(\boldsymbol{R}^m)$. Then $F : \boldsymbol{R}^m \to \boldsymbol{C}$ defined by

$$F(\xi) = (2\pi)^{-2m} \int_{R^m} \int_{R^m} \varphi(\xi - x - y) g(x) \overline{g(-y)} dx dy$$

is continuous and absolutely integrable. From (iv), it follows that $F(0) \geq 0$ (Exercise 2.10). Let

$$\begin{aligned} E(\alpha) &= (2\pi)^{-m} \int e^{-i\langle \alpha, x\rangle} \varphi(x)dx, \\ \Gamma(\alpha) &= (2\pi)^{-m} \int e^{-i\langle \alpha, x\rangle} g(x)dx. \end{aligned}$$

Then

$$| \Gamma(\alpha) |^2 E(\alpha) = (2\pi)^{-m} \int e^{-i\langle \alpha, \xi\rangle} F(\xi)d\xi$$

and thus, since F(.) is continuous,

$$F(\xi) = \int e^{i\langle \xi, \alpha\rangle} | \Gamma(\alpha) |^2 E(\alpha)d\alpha$$

for all ξ. Set $\xi = 0$ to obtain

$$\int | \Gamma(\alpha) |^2 E(\alpha)d\alpha = F(0) \geq 0.$$

Choose g such that $\Gamma(\alpha)$ is continuous and supported in an arbitrarily small neighborhood of some α_0, leading to $E(\alpha_0) \geq 0$. Thus $E \geq 0$ on $\boldsymbol{R}^m$. Since $\varphi(x) = \int e^{i\langle x, \alpha\rangle} E(\alpha)d\alpha, \varphi(0) = 1 = \int E(\alpha)d\alpha$. Hence $\mu(dx) = E(x)dx$ is a probability measure with $\varphi_\mu = \varphi$.

For any φ satisfying (i), (ii), (iv), note that for any positive integrable $\gamma : \boldsymbol{R}^m \to \boldsymbol{R}$ with $\| \gamma \|_1 = 1$, the function $\varphi(x) \int exp(i < x, \alpha >)\gamma(\alpha)d\alpha$ also satisfies (i), (ii), (iv). In particular, take γ to be successively the functions

$$v_n(\alpha) = n^m (2\pi)^{-m/2} exp\left(-\frac{1}{2}\sum_{i=1}^{m}(n\alpha_i)^2\right), n \geq 1, \alpha = [\alpha_1, \cdots, \alpha_m] \in \boldsymbol{R}^m,$$

for which

$$g_n(x) = \int e^{i\langle x, \alpha\rangle} v_n(\alpha)d\alpha = exp\left(-\frac{1}{2}\sum_{i=1}^{m}(x_i/n)^2\right), n \geq 1$$

with $x = [x_1, \cdots, x_n] \in \boldsymbol{R}^m$. For each $n, \varphi g_n$ is absolutely integrable. Hence $\varphi g_n = \varphi_{\mu_n}$ for some $\mu_n \in \boldsymbol{P}(\boldsymbol{R}^m), n \geq 1$, in view of the foregoing. Since $\varphi g_n \to \varphi$ pointwise as $n \to \infty, \mu_n \to \mu$ for some $\mu \in \boldsymbol{P}(\boldsymbol{R}^m)$ with $\varphi_\mu = \varphi$, by virtue of the preceding theorem. □

The concept of characteristic functions fruitfully extends to more general spaces. Let E be a real separable Banach space and E^* its dual. The characteristic function of $\mu \in \boldsymbol{P}(E)$ is the function $\varphi : E^* \to \boldsymbol{C}$ defined by $\varphi(z^*) = E[exp(i\langle Z, z^*\rangle)]$ where $\mathcal{L}(Z) = \mu$ and $\langle ., . \rangle$ is the pairing between E, E^*. We shall extend some of the results above to this case. Let $\boldsymbol{C}$ denote the collection of cylinder sets in E, i.e., sets of the type $\{x \in E \mid (\langle x, z_1\rangle, \cdots, \langle x, z_n\rangle) \in A\}$ for some $n \geq 1, z_i \in E^*, A$ Borel in $\boldsymbol{R}^n$.

Lemma 2.5.2 *$\mathcal{C}$ generates the Borel σ-field of E (implying that the Borel σ-fields of weak and strong topologies on E coincide).*

Proof Let $\{b_n\}$ be countable dense in E. By the Hahn-Banach theorem, we can pick $\{z_n\} \in E^*$ such that $\| z_n \| = 1$ and $\langle b_n, z_n \rangle = \| b_n \|, n \geq 1$. Let $B_1 = \{x \mid \| x \| \leq r\}, B_2 = \bigcap_n \{x \mid \langle x, z_n \rangle \leq r\}$ for some $r > 0$. Then $B_1 \subset B_2$. Let $b \in B_1^c$, implying $\| b \| > r$. Pick a $b_n \in E$ such that

$$\| b - b_n \| < \frac{1}{2}(\| b \| - r).$$

Then

$$\| b_n \| \geq \| b \| - \| b - b_n \| > \frac{1}{2}(\| b \| + r)$$

and

$$| \langle b, z_n \rangle - \| b_n \| | = | \langle b - b_n, z_n \rangle | \leq \| b - b_n \| .$$

Thus

$$\langle b, z_n \rangle \ > \| b_n \| - \frac{1}{2}(\| b \| - r) \ > \frac{1}{2}(\| b \| + r) - \frac{1}{2}(\| b \| - r) = r.$$

Hence $b \in B_2^c$, implying $B_1^c \subset B_2^c$ and therefore $B_1 = B_2$. Thus B_1 is in the σ-field generated by $\mathcal{C}$. A similar argument proves this for any closed ball in E, hence the claim. □

Corollary 2.5.2 *If $\mu, \nu \in \boldsymbol{P}(E)$ satisfy $\varphi_\mu = \varphi_\nu$, then $\mu = \nu$.*

Proof For $n \geq 1, z_1, \cdots, z_n \in E^*, t_1, \cdots, t_n \in \boldsymbol{R}$, we have $\varphi_\mu(\sum_{i=1}^n t_i z_i) = \varphi_\nu(\sum_{i=1}^n t_i z_i)$. By Lemma 2.5.1, μ, ν agree on $\mathcal{C}$. The claim now follows from the above lemma and Theorem 1.2.1. (Exercise 2.11). □

Theorem 2.5.3 *If $\mu_n \to \mu_\infty$ in $\boldsymbol{P}(E), \varphi_{\mu_n} \to \varphi_{\mu_\infty}$ pointwise. Conversely, if $\varphi_{\mu_n} \to \varphi$ pointwise for some $\varphi : E^* \to \boldsymbol{C}$ and $\{\mu_n\}$ is tight, then $\mu_n \to \mu$ for some $\mu \in \boldsymbol{P}(E)$ satisfying $\varphi_\mu = \varphi$.*

The proof is obvious. The definition of positive definiteness extends naturally to E. For $E =$ a real separable Hilbert space H (thus $E^* \approx E$), the following replaces Bochner's theorem.

Theorem 2.5.4 *(Prohorov) A functional $\varphi : H \to \boldsymbol{C}$ is the characteristic function of some $\mu \in \boldsymbol{P}(H)$ if and only if*

(a) $\varphi(0) = 1$ and φ is positive definite,

(b) for every $\epsilon > 0$, there exists a self-adjoint positive definite trace class operator A_ϵ on H such that

$$1 - Re\varphi(x) \leq \langle A_\epsilon x, x \rangle + \ \epsilon, \ x \in H.$$

See [29] for a proof and [29, 38] for further developments along these lines.
For further reading related to this chapter, see [2, 3, 32, 37, 38].

2.6 Additional Exercises

(2.12) Let $\{X_\alpha, \alpha \in I\}$ be $\boldsymbol{R}^m$- valued random variables satisfying the condition : $\sup_\alpha E[\| X_\alpha \|^r] < \infty$ for some $r > 0$. Show that $\{\mathcal{L}(X_\alpha), \alpha \in I\}$ are tight.

(2.13) Let $S_i, i \geq 1$, be Polish spaces and $S = \prod_i S_i$ with the product topology. Show that S is Polish and $\{\mu_\alpha, \alpha \in I\} \subset \boldsymbol{P}(S)$ is tight if and only if $\{\mu_{\alpha n}, \alpha \in I\} \subset \boldsymbol{P}(S_n)$ is tight for each n, where $\mu_{\alpha n}$ is the image of μ_α under the projection $S \to S_n, n \geq 1$.

(2.14) Let $X_n, Y_n, n \geq 1, Z$ be random variables taking values in a Polish space S with a complete metric d. Suppose $X_n \to Z$ in law and $d(X_n, Y_n) \to 0$ in probability. (It is implicit that $\{X_n, Y_n\}$ are defined on a common probability space.) Show that $Y_n \to Z$ in law.

(2.15) For Polish spaces S, S', show that $\mu_n \times \nu_n \to \mu \times \nu$ in $\boldsymbol{P}(S \times S')$ if and only if $\mu_n \to \mu$ in $\boldsymbol{P}(S)$ and $\nu_n \to \nu$ in $\boldsymbol{P}(S')$. (Thus independence is preserved under convergence in law.)

(2.16) Let S be a Polish space endowed with a complete metric d taking values in $[0, 1]$. Show that the metrics q, $\bar{q}$ on $\boldsymbol{P}(S)$ defined above satisfy

$$\bar{q}(\mu, \nu)/2 \;\leq\; q(\mu, \nu) \;\leq\; \sqrt{\bar{q}(\mu, \nu)}$$

for $\mu, \nu \in \boldsymbol{P}(S)$.

(2.17) Let $\varphi : \boldsymbol{R}^m \to \boldsymbol{C}$ be a characteristic function. Prove the following inequalities:

(i) $\mid \varphi(x) - \varphi(y) \mid \leq 2\sqrt{\mid 1 - \varphi(x - y) \mid}$
(ii) $\mid 1 - \varphi(x) \mid \leq \sqrt{2}\sqrt{1 - Re(\varphi(x))}$.

(2.18) A probability distribution μ on $\boldsymbol{R}^d$ is said to be Gaussian, or normal, with mean vector m and covariance matrix Σ (denoted by $N(m, \Sigma)$) if it has a density $p(.)$ with respect to the Lebesgue measure given by

$$p(x) = \frac{1}{(2\pi)^{d/2} \mid det\Sigma \mid^{1/2}} exp(-\frac{1}{2}(x - m)^T \Sigma^{-1}(x - m)).$$

Show that its characteristic function is $\varphi(\nu) = exp(i\langle \nu, m\rangle - \frac{1}{2}\nu^T \Sigma \nu)$.

3

Conditioning and Martingales

3.1 Conditional Expectations

Let $(\Omega, \mathcal{F}, P)$ be a probability space and $A, B \in \mathcal{F}$ events such that $P(B) > 0$. In this section we seek to formalize the intuitive content of the phrase "probability of A given (or, equivalently, "conditioned on") B". The words "given B" imply prior knowledge of the fact that the sample point ω lies in B. It is then natural to consider the reduced probability space $(\Omega_B, \mathcal{F}_B, P_B)$ with $\Omega_B = B, \mathcal{F}_B = \{A \cap \mathrm{B} | \ A \in \mathcal{F}\}$ (the "trace σ-field"), $P_B(C) = P(C)/P(B)$ for $C \in \mathcal{F}_B$, and define the probability of A given B, denoted by $P(A/B)$, as $P_B(A \cap B) = P(AB)/P(B)$. The reader may convince himself of the reasonableness of this procedure by evaluating, say, the probability of two consecutive heads in four tosses of a fair coin given the knowledge that the number of heads was even, using the principle of insufficient reason.

More generally, let $\{B_1, ..., B_m\}$ be a partition of $\Omega, B_i \in \mathcal{F}$ for all i, with $P(B_i) > 0, 1 \leq i \leq m$. Such a partition may be interpreted as the resolution of the sample space that our prior knowledge permits us if we know which of the B_i's the sample point ω belongs to. The updated ("conditional") probability of an event A given this knowledge is $P(AB_i)/P(B_i)$ if ω is in B_i. Thus we may define the probability of A given the partition $\{B_1, ..., B_m\}$ (or, with an eye on the generalization to come, given the

σ-field $\mathcal{G} \subset \mathcal{F}$ generated by this partition) as the random variable

$$P(A/\mathcal{G}) = \sum_{i=1}^{m} (P(AB_i)/P(B_i)) I\{w \in B_i\} \tag{3.1}$$

Two immediate observations are
(i) $P(A/\mathcal{G})$ is $\mathcal{G}$-measurable.
(ii) $P(AC) = \int_C P(A/\mathcal{G})dP$ for all $C \in \mathcal{G}$.

Naturally one would want to extend this notion to a more general sub-σ-field $\mathcal{G}$ of $\mathcal{F}$, interpreted again as an abstraction of the resolution of the sample space which our prior knowledge permits us. Unfortunately, the definition (3.1) depends crucially on $\mathcal{G}$ being generated by a finite (or at best countable) partition $\{B_1, B_2, ...\}$ with $P(B_i) > 0$ for all i. Mercifully, (i) and (ii) do not. Thus we define the conditional probability of an event A given a sub-σ-field $\mathcal{G}$ of $\mathcal{F}$ as the a.s. unique $\mathcal{G}$-measurable random variable $P(A/\mathcal{G})$ which satisfies

$$\int_C P(A/\mathcal{G})dP = P(AC) \text{ for all } C \in \mathcal{G}.$$

This presupposes the existence and uniqueness of such a random variable, but these are easy to establish: If X, X' are two candidates, $X - X'$ is $\mathcal{G}$-measurable and

$$\int_C (X - X')dP = 0 \text{ for all } C \in \mathcal{G}$$

Consider $C = \{X - X' > 0\}$ and $C = \{X - X' < 0\}$ to conclude that $X = X'$ a.s. As for existence, observe that the measure μ on $(\Omega, \mathcal{G})$ defined by $\mu(B) = P(AB)$ for $B \in \mathcal{G}$ is a finite nonnegative measure which is absolutely continuous with respect to P' = the restriction of P to $(\Omega, \mathcal{G})$. Thus we may set $P(A/\mathcal{G}) = d\mu/dP'$. It is easy to verify that when $\mathcal{G}$ is generated by a finite partition $\{B_1, ..., B_m\}$ of Ω with $P(B_i) > 0$, $1 \leq i \leq m$, this definition coincides with (3.1) (Exercise 3.1).

Given a real-valued integrable random variable X on $(\Omega, \mathcal{F}, P)$, one may similarly define the conditional expectation of X given $\mathcal{G}$, denoted by $E[X/\mathcal{G}]$, as the Radon-Nikodym derivative of the finite signed measure μ_X on $(\Omega, \mathcal{G})$ defined by $\mu_X(A) = \int_A X dP, A \in \mathcal{G}$, with respect to P'. Alternatively, it is the a.s. unique $\mathcal{G}$-measurable random variable satisfying

$$\int_B X dP = \int_B E[X/\mathcal{G}]dP, \quad B \in \mathcal{G}.$$

For $X = I_A$, $E[X/\mathcal{G}] = P(A/\mathcal{G})$ as expected. The following lemma lists some elementary properties of conditional expectations which are easily verifiable from its definition (Exercise 3.2). (All equalities/inequalities among random variables are assumed to hold a.s. in what follows.)

Lemma 3.1.1 *Let X_1, X_2 be real integrable random variables on a probability space $(\Omega, \mathcal{F}, P)$ and ξ, $\mathcal{G}$ sub-σ-fields of $\mathcal{F}$ with $\xi \subset \mathcal{G}$. Let $\underline{1}$ = the constant function $\Omega \to \boldsymbol{R}$ identically equal to one. Then*

(i) $E[\underline{1}/\mathcal{G}] = 1$.

(ii) $X_1 \geq X_2 \Rightarrow E[X_1/\mathcal{G}] \geq E[X_2/\mathcal{G}]$.

(iii) If $a,b \in \boldsymbol{R}$, $E[aX_1 + bX_2/\mathcal{G}] = aE[X_1/\mathcal{G}] + bE[X_2/\mathcal{G}]$. *More generally, if Y_1, Y_2 are $\mathcal{G}$-measurable random variables such that $X_iY_i, i = 1, 2$, are integrable, then*

$$E[X_1Y_1 + X_2Y_2/\mathcal{G}] = Y_1E[X_1/\mathcal{G}] + Y_2[X_2/\mathcal{G}].$$

(iv) (filtering property) $E[E[X_1/\mathcal{G}]/\xi] = E[X_1/\xi]$.

Let S be a Polish space. Then (i)-(iii) above suggest that for an S-valued random variable X and bounded measurable $f : S \to \boldsymbol{R}$, $E[f(X)/\mathcal{G}]$ is the integral of f with respect to a random probability measure. The following theorem confirms this intuition.

Theorem 3.1.1 *Let X be an S-valued random variable on $(\Omega, \mathcal{F}, P)$ and $\mathcal{G} \subset \mathcal{F}$ a sub-σ-field. Then there exists an a.s. unique $\mathcal{G}$-measurable map q: $\omega \in \Omega \to q(\omega, dy) \in \boldsymbol{P}(S)$ such that*

$$E[f(X)/\mathcal{G}](\omega) = \int f(y)q(\omega, dy) \quad a.s.$$

for all $f \in C_b(S)$.

Proof Let $C_u(S)$ be as in the proof of Theorem 2.1.1 and $D = \{f_1, f_2, ...\}$ a countable dense set in $C_u(S)$ with $f_1(.) \equiv 1$. Let $N \in \mathcal{G}$ be the set of zero probability outside which $E[f_1(X)/\mathcal{G}] = 1, E[\sum_{i=1}^n a_i f_i(X)/\mathcal{G}] = \sum_{i=1}^n a_i E[f_i(X)/\mathcal{G}]$ for all $n \geq 1, \{f_1, ..., f_n\} \subset D$ and $a_1, ..., a_n$ rational, and (ii) holds for all pairs $X_1, X_2, X_1 \geq X_2$, such that each $X_j, j = 1, 2$, is of the form $\sum_{i=1}^n a_i f_i(X)$ for some $n \geq 1$, $\{f_i\} \subset D$ and $\{a_i\}$ rational. Then for fixed $\omega \notin N$, the map $\sum_{i=1}^n a_i f_i \to E[\sum_{i=1}^n a_i f_i(X)/\mathcal{G}](\omega)$ defines a positive bounded linear functional on the linear span of D over rationals that maps $\underline{1}$ to 1. Since D is dense in $C_u(S)$, it extends uniquely to a positive bounded linear functional on $C_u(S)$ that maps $\underline{1}$ to 1, which in turn may be identified with such a functional on $C(\overline{h(S)}), h$ as in Theorem 2.1.1, via the isometry $C_u(S) \leftrightarrow C(\overline{h(S)})$. By the Riesz theorem, this functional must be of the form $f \to \int f d\mu(\omega)$ for some $\mu(\omega)(= \mu(\omega, dy)) \in \boldsymbol{P}(\overline{h(S)})$.

Extend the map $\omega \to \mu(\omega)$ to entire Ω by setting it equal to a fixed $\mu_0 \in \boldsymbol{P}(\overline{h(S)})$ with $\mu_0(h(S)) = 1$, on N. The $\mathcal{G}$-measurability of the map

$\omega \to \mu(\omega)$ is tantamount to that of the maps $\omega \to \int \mathrm{fd}\mu(\omega)$ for $f \in C(\overline{h(S)})$ and thus $\omega \to \mu(\omega)$ is $\mathcal{G}$-measurable (Exercise 3.3). Now $E[\mu(\omega, h(S))] = E[E[f_1(X)/\mathcal{G}]] = E[f_1(X)] = 1$, implying $\mu(\omega, h(S)) = 1$ a.s. We may drop the "a.s." by modifying $\mu(\omega)$ on a set of zero probability. Identify μ with its restriction to $h(S)$ and let q be its image under h^{-1}. Then $\omega \to q(\omega) = q(w, dy) \in \boldsymbol{P}(S)$ satisfies: $E[f(X)/\mathcal{G}] = \int f dq(\omega)$ a.s. for $f \in D$ and therefore for all $f \in C_b(S)$. Finally, the a.s. uniqueness of $q(.)$ follows from the fact that if $q'(.)$ were another candidate, then necessarily $\int f dq(\omega) = \int f dq'(\omega) = E[f(X)/\mathcal{G}](\omega)$ for all $f \in D$ and all ω outside a zero probability set independent of f. Since D forms a separating class for $\boldsymbol{P}(S), q = q'$ a.s. □.

The $q(.)$ above is called the regular conditional law of X given $\mathcal{G}$.

Corollary 3.1.1 *Let $X_n, n = 1, 2, ..., \infty$ be real integrable random variables on $(\Omega, \mathcal{F}, P)$ and $\mathcal{G} \subset \mathcal{F}$ a sub-σ-field.*

(i) (Conditional Chebyshev's inequality) For $\epsilon > 0$,

$$P(| X_1 |> \epsilon/\mathcal{G}) \leq E[| X_1 | /\mathcal{G}]/\epsilon.$$

(ii) (Conditional Jensen's inequality) Let $\varphi : \boldsymbol{R} \to \boldsymbol{R}$ be convex and $\varphi(X_1)$ integrable. Then

$$E[\varphi(X_1)/\mathcal{G}] \geq \varphi(E[X_1/\mathcal{G}]).$$

(iii) (Conditional monotone convergence theorem) $X_n \uparrow X_\infty$ a.s. implies that $E[X_n/\mathcal{G}] \uparrow E[X_\infty/\mathcal{G}]$ a.s.

(iv) (Conditional dominated convergence theorem) If $X_n \to X_\infty$ a.s. and $| X_n | \leq Y$ a.s., n = 1,2,..., for some integrable random variable Y, then $E[X_n/\mathcal{G}] \to E[X_\infty/\mathcal{G}]$ a.s.

(v) (Conditional Fatou's lemma) If $X_n \geq 0$ a.s. for $n = 1, 2, ...$, then

$$\liminf_{n\to\infty} E[X_n/\mathcal{G}] \geq E[\liminf_{n\to\infty} X_n/\mathcal{G}] \quad a.s.$$

(vi) (Conditional Hölder's inequality) Let $p, q \in (1, \infty)$ with $p^{-1}+q^{-1} = 1$. Suppose that $E[| X_1 |^p], E[| X_2 |^q] < \infty$. Then

$$| E[X_1X_2/\mathcal{G}] |\leq (E[| X_1 |^p /\mathcal{G}])^{1/p}(E[| X_2 |^q /\mathcal{G}])^{1/q}.$$

These are immediate from the preceding theorem.

Corollary 3.1.2 *Let S_1, S_2 be Polish spaces endowed with their Borel σ-fields, $\mu \in \boldsymbol{P}(S_1 \times S_2)$ and $\nu \in \boldsymbol{P}(S_1)$ the image of μ under the projection $S_1 \times S_2 \to S_1$. Then there exists a ν-a.s. unique measurable map $q : S_1 \to \boldsymbol{P}(S_2)$ such that*

$$\mu(dx, dy) = \nu(dx)q(x, dy).$$

Proof In Theorem 3.1.1, take $\Omega = S_1 \times S_2, \mathcal{F} =$ its product σ-field, $P = \mu, \mathcal{G} = \{A \times S_2 \mid A \text{ Borel in } S_1\}$ and $X : \Omega \to S_2$ defined by $X((w_1, \omega_2)) = \omega_2$ for $(\omega_1, \omega_2) \in \Omega$. □

This decomposition of μ is an instance of the so called disintegration of measures [42].

If the sub-σ-field $\mathcal{G}$ of Theorem 3.1.1 is of the form $\mathcal{G} = \sigma(X_1, ..., X_n)$ for some Polish space-valued random variables $X_1, ..., X_n$, then by Theorem 1.1.4, $q(\omega) = F(X_1(\omega), ..., X_n(\omega))$ for a $\boldsymbol{P}(S)$-valued measurable function F. This F is called the regular conditional law of X given $X_1, ..., X_n$. More generally, given a $\mathcal{G}$ as above and an event $A \in \mathcal{F}$ or a real random variable Z on $(\Omega, \mathcal{F}, P), P(A/\mathcal{G}), E[Z/\mathcal{G}]$ are expressible as functions of $X_1, ..., X_n$. These are appropriately called conditional probability of A given $(X_1, ..., X_n)$ and conditional expectation of Z given $(X_1, ..., X_n)$ respectively, and denoted appropriately by $P(A/X_1, ..., X_n), E[Z/X_1, ..., X_n]$ respectively. Pushing this notation a little further, one may consider $\mathcal{G} = \sigma(X_\alpha, \alpha \in I)$ for a family $\{X_\alpha, \alpha \in I\}$ of Polish space-valued random variables and write $P(A/X_\alpha, \alpha \in I), E[Z/X_\alpha, \alpha \in I]$ for $P(A/\mathcal{G}), E[Z/\mathcal{G}]$ resp. By the remarks at the end of Section 1.1, these are expressible as measurable functions of $(X_{i_1}, X_{i_2}, ...)$ for a countable subset $\{i_1, i_2, ...\}$ of I which in general will depend on A or Z as the case may be.

After defining conditional probability and expectation, it is but natural that one should define a conditional variant of the notion of independence. The definition suggests itself: Events $A_1, ..., A_n \in \mathcal{F}$ are said to be conditionally independent given $\mathcal{G}$ (or given $\{X_\alpha, \alpha \in I\}$ if $\mathcal{G} = \sigma(X_\alpha, \alpha \in I)$) if $P(A_1...A_n/\mathcal{G}) = \Pi_{i=1}^n P(A_i/\mathcal{G})$. Sub-$\sigma$-fields $\mathcal{G}_1, ..., \mathcal{G}_n$ of $\mathcal{F}$ are conditionally independent given $\mathcal{G}$ if $A_1, ..., A_n$ are so for any choice of $A_i \in \mathcal{G}_i, 1 \leq i \leq n$. Polish space-valued random variables X_1,...,X_n are conditionally independent given $\mathcal{G}$ if $\sigma(X_1), ..., \sigma(X_n)$ are so. Equivalently, they are conditionally independent given $\mathcal{G}$ if either (i) for $\{A_i\}$ Borel in the appropriate Polish spaces, $P(\cap_{i=1}^n \{X_i \in A_i\}/\mathcal{G}) = \Pi_{i=1}^n P(X_i \in A_i/\mathcal{G})$, or (ii) for bounded continuous real-valued maps $\{f_i\}$ on the appropriate Polish spaces, $E[\Pi_{i=1}^n f_i(X_i)/\mathcal{G}] = \Pi_{i=1}^n E[f_i(X_i)/\mathcal{G}]$, or (iii) the regular conditional law of $[X_1, ..., X_n]$ given $\mathcal{G}$ is a.s. the product of the regular conditional laws of X_i given $\mathcal{G}$ for $1 \leq i \leq n$. Finally, an arbitrary family of events, sub-σ-fields or random variables is conditionally independent given

$\mathcal{G}$ (or, again, given $\{X_\alpha, \alpha \in I\}$ when $\mathcal{G} = \sigma(X_\alpha, \alpha \in I)$) if every finite subfamily thereof is. The next result suggests an alternative interpretation of conditional independence.

Theorem 3.1.2 *Let $(\Omega, \mathcal{F}, P)$ be a probability space and $\mathcal{F}_i, i = 1, 2, 3$, sub-$\sigma$-fields of $\mathcal{F}$. Then $\mathcal{F}_1, \mathcal{F}_3$ are conditionally independent given $\mathcal{F}_2$ if and only if for every integrable $\mathcal{F}_3-$measurable random variable Y,*

$$E[Y/\mathcal{F}_1 \vee \mathcal{F}_2] = E[Y/\mathcal{F}_2]. \tag{3.2}$$

(Equivalently, this is true if and only if for any $\mathcal{F}_1$-measurable integrable random variable $Z, E[Z/\mathcal{F}_2 \vee \mathcal{F}_3] = E[Z/\mathcal{F}_2]$.)

Proof Suppose $\mathcal{F}_1, \mathcal{F}_3$ are conditionally independent given $\mathcal{F}_2$. Let Y be as above. Let $\mathcal{C} = \{A \in \mathcal{F}_1 \bigvee \mathcal{F}_2 \mid \int_A E[Y/\mathcal{F}_1 \bigvee \mathcal{F}_2]dP = \int_A E[Y/\mathcal{F}_2]dP\}$. $\mathcal{C}$ is clearly a $\lambda-$system. Let $A_i \in \mathcal{F}_i, i = 1, 2$. Then

$$\begin{aligned}
E[I_{A_1} I_{A_2} E[Y/\mathcal{F}_1 \vee \mathcal{F}_2]] &= E[E[I_{A_1} I_{A_2} Y/\mathcal{F}_1 \vee \mathcal{F}_2]] \\
&= E[I_{A_1} I_{A_2} Y] \\
&= E[E[I_{A_1} I_{A_2} Y/\mathcal{F}_2]] \\
&= E[I_{A_2} E[I_{A_1} Y/\mathcal{F}_2]] \\
&= E[I_{A_2} E[I_{A_1}/\mathcal{F}_2] E[Y/\mathcal{F}_2]] \\
&= E[I_{A_2} E[I_{A_1} E[Y/\mathcal{F}_2]/\mathcal{F}_2]] \\
&= E[E[I_{A_1} I_{A_2} E[Y/\mathcal{F}_2]/\mathcal{F}_2]] \\
&= E[I_{A_1} I_{A_2} E[Y/\mathcal{F}_2]].
\end{aligned}$$

Thus $A_1 \cap A_2 \in \mathcal{C}$. By Theorem 1.2.1, $\mathcal{C} = \mathcal{F}_1 \bigvee \mathcal{F}_2$, implying (3.2) by virtue of the a.s. uniqueness of conditional expectations. Conversely, let (3.2) hold and let Y_1, Y_3 be arbitrary integrable random variables measurable with respect to $\mathcal{F}_1, \mathcal{F}_3$ respectively such that $Y_1 Y_3$ is integrable. Then

$$\begin{aligned}
E[Y_1 Y_3/\mathcal{F}_2] &= E[E[Y_1 Y_3/\mathcal{F}_1 \vee \mathcal{F}_2]/\mathcal{F}_2] \\
&= E[Y_1 E[Y_3/\mathcal{F}_1 \vee \mathcal{F}_2]/\mathcal{F}_2] \\
&= E[Y_1 E[Y_3/\mathcal{F}_2]/\mathcal{F}_2] \\
&= E[Y_1/\mathcal{F}_2] E[Y_3/\mathcal{F}_2],
\end{aligned}$$

implying that $\mathcal{F}_1, \mathcal{F}_3$ are conditionally independent given $\mathcal{F}_2$. □

3.2 Martingales

Let $(\Omega, \mathcal{F}, P)$ be a probability space. An increasing family $\{\mathcal{F}_n\}$ of sub-σ-fields of $\mathcal{F}$ will be called a filtration. We shall assume throughout that these are P-complete, i.e., contain all P-null events. A sequence $\{X_n\}$ of random variables on $(\Omega, \mathcal{F}, P)$ is said to be adapted to $\{\mathcal{F}_n\}$ if X_n is $\mathcal{F}_n$-measurable for all n. In particular, this is so when $\mathcal{F}_n = \sigma(X_m, m \leq n), n \geq 0$, which is called the natural filtration of $\{X_n\}$. Given a filtration $\{\mathcal{F}_n\}$ and a sequence $\{X_n\}$ of real integrable random variables adapted to it, $(X_n, \mathcal{F}_n), n \geq 0$, is said to be a martingale (resp. a submartingale or a supermartingale) if for $n \geq m, E[X_n/\mathcal{F}_m] = X_m$ (resp. $\geq X_m$ or $\leq X_m$). One often refers to $\{X_n\}$ as a martingale or a sub/supermartingale accordingly when $\{\mathcal{F}_n\}$ is implied or apparent from the context. The term "martingale" comes from gambling. If X_n is interpreted as the reward of a gambler at time n, then a martingale (resp. a sub/supermartingale) corresponds to a fair (resp. more than fair/unfair) game. See [7], Section 5.8 for an interesting account of some applications of martingale theory to gambling.

Examples (i) Let $\{X_n\}$ be independent real integrable random variables with $E[X_n] = 0, n \geq 0$, and let $\mathcal{F}_n = \sigma(X_m, m \leq n), n \geq 0$. Let $S_n = X_0 + \ldots + X_n$. Then $(S_n, \mathcal{F}_n), n \geq 0$, is a martingale. If $E[X_n] \geq 0$ (or ≤ 0) for all n, it is a submartingale (resp. a supermartingale). (Exercise 3.4).

(ii) For $\{X_n\}, \{\mathcal{F}_n\}$ as above, let $f : \boldsymbol{R} \to \boldsymbol{R}$ be a measurable function with $E[\mid f(X_n) \mid] < \infty$ and $a_n = E[f(X_n)] \neq 0$ for all n. Then for $Y_n = (\Pi_{m=0}^n f(X_m))/(\Pi_{m=0}^n a_m), (Y_n, \mathcal{F}_n), n \geq 0$, is a martingale (Exercise 3.5).

(iii) Let $Y \in L_1(\Omega, \mathcal{F}, P), \{\mathcal{F}_n\}$ any filtration and $X_n = E[Y/\mathcal{F}_n], n \geq 0$. Then $(X_n, \mathcal{F}_n), n \geq 0$, is a martingale (Exercise 3.6).

(iv) Let $(X_n, \mathcal{F}_n), n \geq 0$, be a martingale and let $\varphi : \boldsymbol{R} \to \boldsymbol{R}$ a convex map such that $E[\mid \varphi(X_n) \mid] < \infty$ for all n. Then by the conditional Jensen's inequality, $(\varphi(X_n), \mathcal{F}_n), n \geq 0$, is a submartingale. If "convex" is replaced by "concave", it would be a supermartingale. If $(X_n, \mathcal{F}_n), n \geq 0$, were a submartingale (resp. a supermartingale) and φ above convex increasing (resp. concave increasing), then $(\varphi(X_n), \mathcal{F}_n), n \geq 0$, is a submartingale (resp. a supermartingale) (Exercise 3.7).

(v) The negative of a submartingale is a supermartingale and vice versa. A martingale is both a submartingale and a supermartingale.

(vi) Let $\{\mathcal{F}_n\}$ be a filtration and Q another probability measure on $(\Omega, \mathcal{F})$ such that if Q_n, P_n are respectively the restrictions of Q, P to $(\Omega, \mathcal{F}_n)$, then $Q_n \ll P_n, n = 0, 1, 2, \ldots$ Let $\Lambda_n = dQ_n/dP_n, n \geq 0$. Then $(\Lambda_n, \mathcal{F}_n), n \geq 0$, is a martingale (Exercise 3.8).

Another useful concept defined relative to a filtration $\{\mathcal{F}_n\}$ is that of a stopping time. Intuitively, these are random times T such that at any instant n, one can tell whether T has occurred or not purely based on one's knowledge of $\mathcal{F}_n$. To be more precise, a $\{0, 1, 2, \ldots, \infty\}$-valued random variable T is a stopping time (with respect to a filtration $\{\mathcal{F}_n\}$, which is generally implied and not explicitly stated), if for each $n \in \{0, 1, 2, \ldots, \infty\}, \{T \leq$

$n\} \in \mathcal{F}_n$ (equivalently, $\{T = n\} \in \mathcal{F}_n$), where $\mathcal{F}_\infty = \vee_n \mathcal{F}_n$ by convention.

Examples Let $\{X_n\}$ be a sequence of real random variables and $\{\mathcal{F}_n\}$ its natural filtration. Then the following are stopping times (Exercise 3.9). Let $A \subset \boldsymbol{R}$ be Borel.

(1) $T = min\{n \geq 0 \mid X_n \in A\}$ if this set is nonempty, $= \infty$ otherwise. (This is the "first hitting time of" or "first passage time into" A.)

(2) $T = min\{n \geq 0 \mid X_n \notin A\}$ if this set is nonempty, $= \infty$ otherwise. (This is the "first exit time from A".)

(3) $T = min\{n \geq 0 | \mid X_n - X_{n-1} \mid \geq 10\}$ if this set is nonempty, $= \infty$ otherwise.

The following, on the other hand, are not stopping times in general (Exercise 3.10).

(1) $T = sup\{n \geq 0 \mid X_n \in A\}$.

(2) $T = min\{n \geq 0 | \mid X_{n+1} - X_n \mid \geq 10\}$ if the set is nonempty, $= \infty$ otherwise.

Associated with each stopping time T is the stopped σ–field $\mathcal{F}_T$ defined by $\mathcal{F}_T = \{A \in \mathcal{F} \mid A \cap \{T \leq n\} \in \mathcal{F}_n \ \forall n\}$ (equivalently, $= \{A \in \mathcal{F} \mid A \cap \{\tau = n\} \in \mathcal{F}_n \ \forall\, n\}$). It is easy to verify that this is indeed a σ–field (Exercise 3.11). The following properties of stopping times follow from the above definitions (Exercise 3.12).

Lemma 3.2.1 *Let $S, T, T_1, T_2, ...$ be stopping times with respect to a filtration $\{\mathcal{F}_n\}$.*

(i) The following are also stopping times:

$$\sum_{n=1}^{N} T_n, \min(T_1, ..., T_n), \max(T_1, ..., T_n), \sup\{T_i\},$$
$$\inf\{T_i\}, \limsup_{n\to\infty} T_n, \liminf_{n\to\infty} T_n.$$

(ii) If $S \leq T$ a.s., then $\mathcal{F}_S \subset \mathcal{F}_T$.

(iii) $\{T \leq S\}, \{T \geq S\} \in \mathcal{F}_{S\wedge T}$.

(iv) If $A \in \mathcal{F}_T$, then $T_A = TI_A + \infty I_{A^c}$ is a stopping time.

(v) If $Y_n, n = 1, 2, ..., \infty$ is a sequence of random variables adapted to $\{\mathcal{F}_n\}$, then Y_T is $\mathcal{F}_T$-measurable. (In particular, T is $\mathcal{F}_T$-measurable.)

(vi) For $\{Y_n\}$ as above, $E[Y_T/\mathcal{F}_S]$ is $\mathcal{F}_{T\wedge S}$-measurable when defined.

The following is among the most important results of martingale theory.

Theorem 3.2.1 *(Doob's optional sampling theorem) Let $\{X_n\}$ be $\{\mathcal{F}_n\}$-adapted integrable real random variables. The following are equivalent:*

(i) $(X_n, \mathcal{F}_n), n \geq 0$, is a submartingale.

(ii) If T, S are bounded stopping times with $T \geq S$ a.s., then $E[X_S] \leq E[X_T]$.

(iii) If S is a stopping time and T a bounded stopping time, then

$$E[X_T/\mathcal{F}_S] \geq X_{S\wedge T}.$$

Proof (i) $\Rightarrow$ (ii) Let $S \leq T \leq n$. Then

$$\begin{aligned}
X_T - X_S &= \sum_{k=1}^{n}\sum_{j=0}^{k-1} I\{T=k\}I\{S=j\}(X_k - X_j) \\
&= \sum_{k=1}^{n} I\{T=k\}I\{S<k\}X_k - \sum_{j=0}^{n-1} I\{S=j\}I\{T>j\}X_j \\
&= \sum_{k=1}^{n} I\{T=k\}I\{S<k\}X_k - \sum_{k=1}^{n} I\{S=k-1\}I\{T\geq k\}X_{k-1} \\
&= \sum_{k=1}^{n} I\{S<k\}I\{T\geq k\}X_k - \sum_{k=1}^{n} I\{S<k\}I\{T\geq k\}X_{k-1}
\end{aligned}$$

(after adding and subtracting $\sum_{k=1}^{n} I\{S<k-1\}I\{T\geq k\}X_{k-1}$). Thus

$$\begin{aligned}
E[X_T - X_S] &= \sum_{k=1}^{n} E[I\{S<k\}I\{T\geq k\}[X_k - X_{k-1}]] \\
&= \sum_{k=1}^{n} E[I\{S<k\}I\{T\geq k\}E[X_k - X_{k-1}/\mathcal{F}_{k-1}]] \geq 0.
\end{aligned}$$

(ii) $\Rightarrow$ (iii) Let $R = S\wedge T, A \in \mathcal{F}_R, U = R_A \wedge n, V = T_A \wedge n$ where R_A, T_A are defined as in Lemma 3.2.1(iv). Then U, V, R_A, T_A are stopping times and $V \geq U$ a.s., implying $E[X_U] \leq E[X_V]$. Hence for n satisfying $n \geq T$,

$$\int_A X_R + \int_{A^c} X_n \leq \int_A X_T + \int_{A^c} X_n$$

for all $A \in \mathcal{F}_{S\wedge T}$. Thus $E[X_T/\mathcal{F}_{S\wedge T}] \geq X_{S\wedge T}$ a.s. But $E[X_T/\mathcal{F}_S]$ is $\mathcal{F}_{S\wedge T}$-measurable and thus $E[X_T/\mathcal{F}_S] = E[X_T/\mathcal{F}_{S\wedge T}]$. This proves (iii).

(iii) $\Rightarrow$ (i) is easy. □

Corollary 3.2.1 *If $(X_n, \mathcal{F}_n), n \geq 0$, is a submartingale and $\{T_n\}$ an increasing sequence of bounded stopping times, then $(X_{T_n}, \mathcal{F}_{T_n}), n \geq 0$, is a submartingale.*

This is immediate from the above theorem. Thus the submartingale property is preserved under the "optional sampling" of the original sequence of random variables and the associated filtration.

Corollary 3.2.2 *(Doob's maximal and minimal inequalities) Let $(X_n, \mathcal{F}_n), n \geq 0$, be a submartingale. Then for all $\lambda > 0$,*

$$\begin{aligned}
\lambda P(\max_{n \leq N} X_n \geq \lambda) &\leq \int_{\{\max_{n\leq N} X_n \geq \lambda\}} X_N \, dP \leq E[X_N^+], \\
\lambda P(\min_{n \leq N} X_n \leq -\lambda) &\leq -E[X_0] + \int_{\{\min_{n\leq N} X_n \geq -\lambda\}} X_N dP \\
&\leq -E[X_0] + E[X_N^+].
\end{aligned}$$

Proof Let $T = (min\{n \mid X_n \geq \lambda\}) \wedge N$. Then T is a stopping time and

$$E[X_N] \geq E[X_T] \geq \lambda \int_{\{\max_{n\leq N} X_n \geq \lambda\}} dP + \int_{\{\max_{n\leq N} X_n < \lambda\}} X_N dP.$$

Therefore

$$\begin{aligned}
\lambda P(\max_{n \leq N} X_n \geq \lambda) \leq E[X_N] - \int_{\{\max_{n\leq N} X_n < \lambda\}} X_N dP. \\
= \int_{\{\max_{n\leq N} X_n \geq \lambda\}} X_N dP \leq E[X_N^+].
\end{aligned}$$

The second claim follows along similar lines by taking $T = (min\{n \mid X_n \leq -\lambda\}) \wedge N$ (Exercise 3.13). □

Corollary 3.2.3 *Let $(X_n, \mathcal{F}_n)$ be a martingale or a nonnegative submartingale. Let $E[\mid X_n \mid^p] < \infty$ for some $p \in (1, \infty)$. Then*

$$E[\mid \max_{n \leq N} X_n \mid^p] \leq \left(\frac{p}{p-1}\right)^p E[\mid X_N \mid^p].$$

For $p = 1$, one has

$$E[\mid \max_{n \leq N} X_n \mid] \leq \frac{e}{e-1}(1 + E[\mid X_N \mid ln^+ \mid X_N \mid])$$

where $ln^+ x = max(0, ln x)$.

Proof Without any loss of generality, let $X_n \geq 0$ a.s. for all n. Then

$$\begin{aligned}
E[(\max_{n\leq N} X_n)^p] &= p\int_0^\infty \lambda^{p-1}P(\max_{n\leq N} X_n \geq \lambda)d\lambda \\
&\leq p\int_0^\infty \lambda^{p-2}E[X_N I\{\max_{n\leq N} X_n \geq \lambda\}]d\lambda \\
&= p\int_0^\infty \lambda^{p-2}d\lambda \int_0^\infty P(X_N \geq \mu, \max_{n\leq N} X_n \geq \lambda)d\mu \\
&= \frac{p}{p-1}\int_0^\infty E[(\max_{n\leq N} X_N)^{p-1}I\{X_N \geq \mu\}]d\mu \\
&= \frac{p}{p-1}E[(\max_{n\leq N} X_n)^{p-1}X_N] \\
&\leq \frac{p}{p-1}E[(\max_{n\leq N} X_N)^p]^{1-1/p}E[(X_N)^p]^{1/p}.
\end{aligned}$$

Divide both sides by $E[(\max_{n\leq N} X_N)^p]^{1-1/p}$ and take the p-th power to conclude the proof of the first inequality. For the second, we have

$$\begin{aligned}
E[\max_{n\leq N} X_n] - 1 &\leq E[(\max_{n\leq N} X_n - 1)^+] \\
&= \int_0^\infty P(\max_{n\leq N} X_n - 1 \geq \lambda)d\lambda \\
&\leq \int_0^\infty (\lambda+1)^{-1}\int_{\{\max_{n\leq N} X_n \geq \lambda+1\}} X_N dP d\lambda \\
&\quad \text{(by Doob's maximal inequality)} \\
&\leq E\left[X_N \int_0^{\max_{n\leq N} X_n - 1} \frac{d\lambda}{\lambda+1}\right] \\
&= E[X_N ln(\max_{n\leq N} X_n)].
\end{aligned}$$

It is easily verified that for $a \geq 0, b > 0, alnb \leq aln^+a + be^{-1}$ (Exercise 3.14). Thus

$$E[\max_{n\leq N} X_n] - 1 \leq E[X_N ln^+ X_N] + e^{-1}E[\max_{n\leq N} X_n].$$

The claim follows. □

Let $(X_n, \mathcal{F}_n), n \geq 0$, be a submartingale and (a, b) a nonempty interval. Let $N \geq 1$. Define the stopping times

$$\begin{aligned}
T_0 &= 0, \\
T_1 &= min\{0 \leq n \leq N \mid X_n \leq a\}, \\
T_2 &= min\{T_1 < n \leq N \mid X_n \geq b\},
\end{aligned}$$

and for $m = 2, 3, \ldots,$

$$T_{2m-1} = min\{T_{2m-2} < n \leq N \mid X_n \leq a\},$$

$$T_{2m} = min\{T_{2m-1} < n \leq N \mid X_n \geq b\}.$$

If $\min_{n\leq N} X_n > a$, then $T_1 = N$ and $T_2, T_3, ...,$ are undefined. A similar convention is followed for $T_2, T_3, ...$ The number of upcrossings of (a, b) up to time N is the random variable $\beta_N(a, b)$ = the maximum m for which T_{2m} is defined (= the number of times between 0 and N that $\{X_n\}$ crosses the strip (a, b) from below a to above b).

Corollary 3.2.4 *(Doob's upcrossing inequality)*

$$E[\beta_N(a, b)] \leq \frac{E[(X_N - a)^+]}{(b-a)} \leq \frac{E[X_N^+] + \mid a \mid}{(b-a)}$$

Proof Since $\beta_N(a, b)$ for the sequence $\{X_n\}$ is same as $\beta_N(0, b-a)$ for the sequence $\{(X_n - a)^+\}$, we may take $a = 0$ and $X_n \geq 0$ without any loss of generality. Also, we may and do take $X_0 = 0$. For $i = 1, 2, ...,$ let $\alpha_i = I\{T_m < i \leq T_{m+1}$ for some odd $m\}$. Then

$$b\beta_N(a, b) \leq \sum_{i=1}^{N} \alpha_i(X_i - X_{i-1})$$

and

$$\{\alpha_i = 1\} = \bigcup_{m\ odd} (\{T_m < i\} \backslash \{T_{m+1} < i\}) \in \mathcal{F}_{i-1}, i \geq 1.$$

Thus

$$\begin{aligned} bE[\beta_N(a, b)] &\leq E[\sum_{i=1}^{N} \alpha_i(X_i - X_{i-1})] \\ &= \sum_{i=1}^{N} \int_{\{\alpha_i = 1\}} (X_i - X_{i-1}) dP \\ &= \sum_{i=1}^{N} \int_{\{\alpha_i = 1\}} (E[X_i/\mathcal{F}_{i-1}] - X_{i-1}) dP \\ &\leq \sum_{i=1}^{N} \int_{\Omega} (E[X_i/\mathcal{F}_{i-1}] - X_{i-1}) dP \\ &= E[X_N]. \end{aligned}$$

This completes the proof. □

This inequality will play a key role in the next section. We conclude this section with yet another important result due to Doob.

Theorem 3.2.2 *(Doob decomposition) A submartingale $(X_n, \mathcal{F}_n), n \geq 0$, can be written as $X_n = M_n + A_n, n \geq 0$, where $(M_n, \mathcal{F}_n), n \geq 0$, is a zero mean martingale and $\{A_n\}$ is an increasing process (i.e., $A_i \leq A_{i+1}$ a.s. for all i) adapted to $\{\mathcal{F}_{n-1}\}$; where $\mathcal{F}_{-1} = \{\phi, \Omega\}$ by convention. This decomposition is a.s. unique.*

Proof Take $M_n = \sum_{i=0}^{n}(X_i - E[X_i/\mathcal{F}_{i-1}])$, $A_n = \sum_{i=1}^{n}(E[X_i/\mathcal{F}_{i-1}] - X_{i-1}) + E[X_0], n \geq 0$. This clearly meets the requirement. If $(M'_n, A'_n), n \geq 0$, is another pair which does, then $M_n + A_n = M'_n + A'_n$. Thus $(A_n - A'_n, \mathcal{F}_n), n \geq 0$, is a martingale. Thus $E[A_n - A'_n/\mathcal{F}_{n-1}] = A_{n-1} - A'_{n-1}, n \geq 1$. Since A_n, A'_n are $\mathcal{F}_{n-1}$-measurable, this implies $A_n - A'_n = A_{n-1} - A'_{n-1} = ... = A_0 - A'_0$. But A_0, A'_0 are $\mathcal{F}_{-1}$-measurable and hence a.s. constant. Thus $A_0 = E[X_0 - M_0] = E[X_0] = A'_0$. Thus $A_n = A'_n$ for all n. Hence $M_n = M'_n$ a.s. for all n. □

3.3 Convergence Theorems

Many important uses of martingale theory in applications are rooted in the many convergence theorems therein. We shall consider several of these here. The basic "submartingale convergence theorem" is the following:

Theorem 3.3.1 *Let $(X_n, \mathcal{F}_n), n \geq 0$, be a submartingale satisfying $\sup_n E[X_n^+] < \infty$. Then $\lim_{n\to\infty} X_n = X_\infty$ exists a.s.*

Proof Let $\overline{X} = \limsup_{n\to\infty} X_n, \underline{X} = \liminf_{n\to\infty} X_n$. Suppose that $P(\overline{X} > \underline{X}) > 0$. Since $\{\overline{X} > \underline{X}\} = \bigcup_{a,b\ rational}\{\overline{X} > b > a > \underline{X}\}$, there exist rational $b > a$ such that $P(\overline{X} > b > a > \underline{X}) > 0$. But

$$E[\beta_N(a,b)] \leq (E[X_N^+] + \mid a \mid)/(b-a).$$

Letting $N \to \infty, \beta_\infty(a,b) = \lim_{N\to\infty} \beta_N(a,b)$ satisfies

$$E[\beta_\infty(a,b)] \leq (\sup_n E[X_n^+] + \mid a \mid)/(b-a) < \infty.$$

Thus $\beta_\infty(a,b) < \infty$ a.s., contradicting $P(\overline{X} > b > a > \underline{X}) > 0$. Thus $P(\overline{X} > b > a > \underline{X}) = 0$ for all rational a, b with $b > a$, implying $P(\overline{X} > \underline{X}) = 0$. Thus $\{X_n\}$ must a.s. converge in $[-\infty, \infty]$. If $P(X_n \to \infty) > 0$, Fatou's lemma gives

$$\infty > \limsup_{n\to\infty} E[X_n^+] \geq E[\liminf_{n\to\infty} X_n^+] = \infty,$$

a contradiction. Thus $P(X_n \to \infty) = 0$. Now $E[X_n] = E[X_n^+] - E[X_n^-] \geq E[X_0]$ for all n. Thus

$$\sup_n E[X_n^-] \leq \sup_n E[X_n^+] + E[\mid X_0 \mid] < \infty.$$

Argue as above to conclude $P(X_n \to -\infty) = 0$. The claim follows. □

The condition $\sup_n E[X_n^+] < \infty$ is trivially satisfied by nonpositive submartingales. Thus nonnegative martingales and supermartingales converge a.s.

Corollary 3.3.1 *Let $\{X_n\}$ above be a martingale or a nonnegative submartingale and satisfy $sup_n E[| X_n |^p] < \infty$ for some $p \in (1, \infty)$. Then $X_n \to X_\infty$ in L_p as well.*

Proof Letting $N \to \infty$ in Corollary 3.2.3, we have

$$E[| \sup_n X_n |^p] \leq \left(\frac{p}{p-1}\right)^p \sup_n E[| X_n |^p].$$

By Theorem 3.3.1, $X_n \to X_\infty$ a.s. for some random variable X_∞. Now

$$| X_n - X_\infty |^p \leq 2^p(| X_n |^p + | X_\infty |^p) \leq 2^{p+1} \sup_n | X_n |^p .$$

The claim follows from the dominated convergence theorem. □

The following example shows that the claim is false for $p = 1$. Let $\Omega = [0, 1]$, $\mathcal{F} =$ its Borel σ-field, $P =$ the Lebesgue measure. Let

$$X_n(\omega) = nI\{\omega \in (0, 1/n)\}, n \geq 1,$$

and $\{\mathcal{F}_n\}$ its natural filtration. Then $(X_n, \mathcal{F}_n), n \geq 1$, is a martingale with $E[X_n] = 1$ for all n (Exercise 3.15). But $X_n \to 0$ a.s. The case $p = 1$ is rather special and we investigate it next. Call a martingale $(X_n, \mathcal{F}_n), n \geq 0$, regular if there exists an integrable random variable η such that $X_n = E[\eta/\mathcal{F}_n]$ for all n.

Theorem 3.3.2 *Let $(X_n, \mathcal{F}_n), n \geq 0$, be a martingale. Then the following are equivalent:*
(i) $(X_n, \mathcal{F}_n), n \geq 0$, is regular.
(ii) $\{X_n\}$ is u.i.
(iii) There exists an integrable random variable X_∞ s.t. $E[| X_n - X_\infty |] \to 0$.
(iv) $\sup_n E[| X_n |] < \infty$ and $X_\infty = \lim_{n\to\infty} X_n$ (which exists by the preceding theorem) satisfies: $X_n = E[X_\infty/\mathcal{F}_n]$ for all n.

Proof (i) $\Rightarrow$ (ii) Let η be as above then

$$\sup_n E[| X_n |] \leq E[| \eta |] < \infty.$$

Also, for $b, c > 0$,

$$\begin{aligned}
&\int_{\{|X_n|\geq c\}} | X_n | \, dP \\
&\leq \int_{\{|X_n|\geq c\}} | \eta | \, dP \\
&\leq \int_{\{|X_n|\geq c\}\cap\{|\eta|\geq b\}} | \eta | \, dP + \int_{\{|X_n|\geq c\}\cap\{|\eta|< b\}} | \eta | \, dP \\
&\leq bP(| X_n |\geq c) + \int_{\{|\eta|\geq b\}} | \eta | \, dP \\
&\leq \frac{b}{c} E[| X_n |] + \int_{\{|\eta|\geq b\}} | \eta | \, dP.
\end{aligned}$$

So

$$\begin{aligned}
\sup_n \int_{\{|X_n|\geq c\}} | X_n | \, dP &\leq \frac{b}{c} E[| \eta |] + \int_{\{|\eta|\geq b\}} | \eta | \, dP \\
&\to \int_{\{|\eta|\geq b\}} | \eta | \, dP
\end{aligned}$$

as $c \to \infty$. Since b was arbitrary, the claim follows.
(ii) $\Rightarrow$ (iii) follows from Theorem 3.3.1 and the fact that a.s. convergence and uniform integrability imply L_1- convergence.
(iii) $\Rightarrow$ (iv). Clearly, $\sup_n E[| X_n |] < \infty$. Thus $Y = \lim_{n\to\infty} X_n$ exists a.s. and we must have $Y = X_\infty$ a.s. Now $E[X_n/\mathcal{F}_m] \to E[X_\infty/\mathcal{F}_m]$ in $L_1(\Omega, \mathcal{F}, P)$ as $m \leq n \to \infty$. But $E[X_n/\mathcal{F}_m] = X_m$ for all $n \geq m$. Thus $X_m = E[X_\infty/\mathcal{F}_m]$.
(iv) $\Rightarrow$ (i) Take $\eta = X_\infty$. □

Another important result in this context is the "Krickeberg decomposition" described in the next theorem.

Theorem 3.3.3 *Let $(X_n, \mathcal{F}_n), n \geq 0$, be a martingale. Then one has $\sup_n E[| X_n |] < \infty$ if and only if there exist nonnegative $\{\mathcal{F}_n\}$-martingales $\{Y_n\}$ and $\{Z_n\}$ such that $X_n = Y_n - Z_n, n \geq 0$.*

Proof Suppose such a decomposition holds. Then

$$\begin{aligned}
\sup_n E[| X_n |] &\leq \sup_n E[Y_n] + \sup_n E[Z_n] \\
&= E[Y_0] + E[Z_0] < \infty.
\end{aligned}$$

Conversely, let $\sup E[| X_n |] < \infty$. Then $(X_n^+, \mathcal{F}_n), n \geq 0$, is a submartingale and for $k \geq n \geq m$, $E[X_k^+/\mathcal{F}_n] \geq X_n^+$, implying $E[X_k^+/\mathcal{F}_m] \geq E[X_n^+/\mathcal{F}_m]$. Then $Y_m = \lim_{k\to\infty} E[X_k^+/\mathcal{F}_m]$ exists a.s., with the monotone convergence theorem leading to

$$E[Y_n] = \lim_{k\to\infty} E[E[X_k^+/\mathcal{F}_n]]$$

$$\begin{aligned} &= \lim_{k\to\infty} E[X_k^+] \\ &\leq \sup_m E[| X_m |] < \infty. \end{aligned}$$

By the conditional monotone convergence theorem,

$$\begin{aligned} E[Y_n/\mathcal{F}_{n-1}] &= \lim_{k\to\infty} E[E[X_k^+/\mathcal{F}_n]/\mathcal{F}_{n-1}] \\ &= \lim_{k\to\infty} E[X_k^+/\mathcal{F}_{n-1}] = Y_{n-1} \quad \text{a.s.}, n = 1, 2, \ldots . \end{aligned}$$

Thus $(Y_n, \mathcal{F}_n)$ is a nonnegative martingale. Also, $X_n \leq X_n^+ \leq Y_n$ for all n. Thus $Z_n = Y_n - X_n, n \geq 0$, is also a nonnegative $\{\mathcal{F}_n\}$-martingale. □

This decomposition is not unique. For example, use of $| X_n |$ in place of X_n^+ in the foregoing leads to another, as also the addition of a positive constant to both $\{Y_n\}$ and $\{Z_n\}$. The specific pair $\{Y_n\}, \{Z_n\}$ above, however, is minimal in the following sense.

Corollary 3.3.2 *If $\{X_n\}$ above satisfies $X_n = Y_n' - Z_n', n \geq 0$, where $\{Y_n\}, \{Z_n\}$ are nonnegative $\{\mathcal{F}_n\}$-martingales, then $Y_n' \geq Y_n, Z_n' \geq Z_n$ for all n.*

Proof Clearly $Y_n' \geq X_n^+$ for all n. Let $n \geq m$. Then

$$Y_m' = E[Y_n'/\mathcal{F}_m] \geq E[X_n^+/\mathcal{F}_m]$$

and therefore

$$Y_m' \geq \lim_{n\to\infty} E[X_n^+/\mathcal{F}_m] = Y_m.$$

The claim follows. □

Another interesting class of martingales is that of square-integrable martingales, that is, martingales $(X_n, \mathcal{F}_n), n \geq 0$, satisfying $E[X_n^2] < \infty$ for all n. For such $\{X_n\}, (X_n^2, \mathcal{F}_n)$ is a submartingale and thus has a Doob decomposition $X_n^2 = M_n + A_n, n \geq 0$, where $(M_n, \mathcal{F}_n), n \geq 0$, is a zero mean martingale and $\{A_n\}$ is an $\{\mathcal{F}_{n-1}\}$-adapted increasing process called the "quadratic variation process" of $\{X_n\}$. One easily verifies that the latter in fact will be

$$\begin{aligned} A_n &= \sum_{m=1}^{n} (E[X_m^2/\mathcal{F}_{m-1}] - X_{m-1}^2) + E[X_0^2] \\ &= \sum_{m=0}^{n-1} E[(X_{m+1} - X_m)^2/\mathcal{F}_m] + E[X_0^2], \ n \geq 0. \end{aligned}$$

This process plays a key role in the analysis of the asymptotic behaviour of $\{X_n\}$, a glimpse of which can be had from the next theorem. We start with two technical lemmas.

Lemma 3.3.1 *In the above, let T be an $\{\mathcal{F}_n\}$-stopping time. Then $(X_{T\wedge n}, \mathcal{F}_{T\wedge n}), n \geq 0$, is a square-integrable martingale whose quadratic variation process is $\{A_{T\wedge n}\}$.*

This follows easily from the optional sampling theorem (Exercise 3.16).

Lemma 3.3.2 *(Kronecker's lemma) Let $\{x_k\} \subset \mathbf{R}, \{a_k\} \subset (0,\infty)$ be such that $a_k \uparrow \infty$ and $|\sum_n (x_n/a_n)| < \infty$. Then $(\sum_{m=1}^n x_m)/a_n \to 0$.*

Proof Let $b_n = \sum_{j=1}^n (x_j/a_j), n = 1, 2, ..., \infty, b_0 = 0 = a_0$. Then $x_n = a_n(b_n - b_{n-1})$ for all n and

$$\frac{1}{a_n}\sum_{j=1}^n x_j = \frac{1}{a_n}\sum_{j=1}^n a_j(b_j - b_{j-1}) = b_n - \frac{1}{a_n}\sum_{j=0}^{n-1} b_j(a_{j+1} - a_j).$$

Since $a_{j+1} - a_j \geq 0$ for all j, $[\sum_{j=0}^{n-1}(a_{j+1} - a_j)]/a_n = 1$ and $b_n \to b_\infty$, we have

$$\frac{1}{a_n}\sum_{j=1}^n x_j \to b_\infty - b_\infty = 0.$$ □

Theorem 3.3.4 *For $\{X_n\}, \{\mathcal{F}_n\}, \{A_n\}$ as above, let $A_\infty = \lim_{n\to\infty} A_n$. Then $\{X_n\}$ converges a.s. on $\{A_\infty < \infty\}$ and $X_n = o(f(A_n))$ on $\{A_\infty = \infty\}$ for every increasing $f : [0,\infty) \to [0,\infty)$ satisfying $\int_0^\infty (1 + f(t))^{-2} dt < \infty$ (e.g., $f(t) = t$).*

Proof For $a > 0$, let $T_a = min\{n \mid A_{n+1} > a^2\}$ if $A_\infty > a^2, = \infty$ otherwise. Then T_a is a stopping time and

$$E[\| X_{T_a\wedge n} \|] \leq E[X_{T_a\wedge n}^2]^{1/2} = E[A_{T_a\wedge n}]^{1/2} \leq a.$$

Thus $\{X_{T_a\wedge n}\}$ converges a.s., that is, $\{X_n\}$ converges a.s. on $\{T_a = \infty\} = \{A_\infty \leq a\}$. Let $a = 1, 2, ...,$ successively to conclude that $\{X_n\}$ converges a.s. on $\{A_\infty < \infty\}$. To prove the second claim, first observe that $\lim_{t\to\infty} f(t) = \infty$. Let

$$Z_n = \sum_{m<n}(X_{m+1} - X_m)/(1 + f(A_{m+1})), n \geq 0.$$

Then $(Z_n, \mathcal{F}_n), n \geq 0$, is a square-integrable martingale with zero mean. Let $\{B_n\}$ be its quadratic variation process. Then for every $n \geq 0$,

$$\begin{aligned} B_{n+1} - B_n &= E[Z_{n+1}^2 - Z_n^2/\mathcal{F}_n] \\ &= E[(Z_{n+1} - Z_n)^2/\mathcal{F}_n] \\ &= E[(X_{n+1} - X_n)^2/\mathcal{F}_n]/(1 + f(A_{n+1}))^2 \end{aligned}$$

$$\begin{aligned} &= E[X_{n+1}^2 - X_n^2/\mathcal{F}_n]/(1+f(A_{n+1}))^2 \\ &= (A_{n+1} - A_n)/(1+f(A_{n+1}))^2. \end{aligned}$$

But

$$\begin{aligned} \sum_{n=1}^{N} \frac{A_{n+1} - A_n}{(1+f(A_{n+1}))^2} &\leq \sum_{n=1}^{N} \int_{A_n}^{A_{n+1}} \frac{dt}{(1+f(t))^2} \\ &\leq \int_0^{\infty} \frac{dt}{(1+f(t))^2} < \infty. \end{aligned}$$

By the first part of this theorem, $\{Z_n\}$ converges a.s. The claim now follows from the preceding lemma and the fact that $\{f(A_n) \to \infty\} = \{A_\infty = \infty\}$.□

The next result suggests that a martingale will oscillate violently when it does not converge.

Theorem 3.3.5 *Let* $(X_n, \mathcal{F}_n)$, $n \geq 0$, *be a martingale which satisfies the condition* $E[\sup_n \mid X_{n+1} - X_n \mid] < \infty$. *Let* $A_1 = \{X_n$ *converges*$\}$, $A_2 = \{\limsup_{n\to\infty} X_n = -\liminf_{n\to\infty} X_n = \infty\}$. *Then* $P(A_1 \cup A_2) = 1$.

Proof Let $a > 0$ and $T = min\{n \geq 1 \mid X_n \geq a\}$ when this set is nonempty and $= \infty$ otherwise. Then $(X_{T\wedge n}, \mathcal{F}_{T\wedge n}), n \geq 0$, is a martingale which satisfies $\sup_n E[X^+_{T\wedge n}] \leq a + E[\sup_n \mid X_{n+1} - X_n \|] < \infty$. Thus $\{X_{T\wedge n}\}$ converges a.s. That is, $\{X_n\}$ converges a.s. on $\{T = \infty\} = \{\sup_n X_n \leq a\}$. Taking $a = 1, 2, \ldots$, successively, it follows that $\{X_n\}$ converges a.s. on $\{\limsup_{n\to\infty} X_n < \infty\}$. A symmetric argument applied to $(-X_n, \mathcal{F}_n), n \geq 0$, shows that $\{X_n\}$ converges a.s. on $\{\liminf_{n\to\infty} X_n > -\infty\}$. □

There are several other convergence theorems related to the foregoing which find wide applications in, for example, convergence analysis of stochastic approximation and recursive parameter estimation algorithms. As an example, consider the following result for "almost supermartingales".

Theorem 3.3.6 *Let* $\{X_n\}, \{\beta_n\}, \{Y_n\}$ *be sequences of nonnegative random variables adapted to a filtration* $\{\mathcal{F}_n\}$ *and satisfying*

$$E[X_{n+1}/\mathcal{F}_n] \leq (1+\beta_n)X_n + Y_n, \ n \geq 0.$$

Then $\{X_n\}$ *converges a.s. on the set* $\{\sum_n \beta_n < \infty, \sum_n Y_n < \infty\}$.

Proof Define $X'_n = X_n/\Pi_{i=0}^{n-1}(1+\beta_i), Y'_n = Y_n/\Pi_{i=0}^{n}(1+\beta_i), n \geq 1$, with $X'_0 = X_0$ and $Y'_0 = Y_0$. Then $E[X'_{n+1}/\mathcal{F}_n] \leq X'_n + Y'_n, n \geq 0$. Define $U_n = X'_n - \sum_{i=0}^{n-1} Y'_i, n \geq 1, U_0 = X_0$. Then $E[U_{n+1}/\mathcal{F}_n] \leq U_n, n \geq 0$. Let $a > 0$ and $\nu_a = min\{n \geq 0 \mid \sum_{m=0}^{n} Y'_m > a\}$ if this set is nonempty and $= \infty$ otherwise. It follows that $(a + U_n)I_{\{\nu_a > n\}}, n \geq 0$, is a nonnegative $\{\mathcal{F}_n\}$-supermartingale and hence converges a.s. Thus $\{U_n\}$ converges a.s.

on the set $\{\nu_a = \infty\} = \{\sum_n Y_n' < a\}$. Since a was arbitrary, it follows that $\{U_n\}$ converges a.s. on $\{\sum_n Y_n' < \infty\}$. Thus $\{X_n'\}$ converges a.s. on $\{\sum_n Y_n' < \infty\}$ and therefore on $\{\sum_n Y_n < \infty\}$ in view of the inequality $0 \le Y_n' \le Y_n, n \ge 0$. Hence $\{X_n\}$ converges a.s. on $\{\sum_n Y_n < \infty, \Pi_{n=0}^{\infty}(1+\beta_n) < \infty\}$. Using the inequality $1 + x \le e^x$, one has $\Pi_{n=0}^{\infty}(1+\beta_n) < \infty$ on the set $\{\sum_n \beta_n < \infty\}$. This completes the proof. □

So far we considered an increasing family of sub-σ-fields $\{\mathcal{F}_n\}$ and processes adapted to it. One may similarly consider a decreasing family of sub-σ-fields $\{\mathcal{B}_n\}$, that is, a "backward filtration", and processes adapted to it. Let $\{X_n\}$ be real integrable random variables adapted to a backward filtration $\{\mathcal{B}_n\}$. It is called a reversed martingale (resp. a reversed submartingale or supermartingale) if $E[X_n/\mathcal{B}_{n+1}] = X_{n+1}$ (resp. $\ge$ or $\le X_{n+1}$).

Theorem 3.3.7 *Let $(X_n, \mathcal{B}_n), n \ge 0$, be a reversed submartingale. Then $\{X_n\}$ converges in $[-\infty, \infty)$ a.s. If it satisfies $\lim_{n\to\infty} E[X_n] = K > -\infty$, then the convergence is also in L_1. Furthermore, if $X_n \ge 0$ a.s. for all n and $E[\mid X_0 \mid^p] < \infty$ for some $p \in (1, \infty)$, then the convergence is also in L_p.*

Proof For $n \ge 1, \mathcal{B}_n, \mathcal{B}_{n-1}, ..., \mathcal{B}_0$ is an increasing family of sub-σ-fields and $X_n, X_{n-1}, ..., X_0$ a usual submartingale with respect to this family. For $b > a$, let $\beta_N(a,b)$ = the number of upcrossings of (a,b) by $X_n, ..., X_0$ = the number of "downcrossings" of (a,b) by $X_0, ..., X_n$. By the upcrossing inequality,

$$E[\beta_N(a,b)] \le E[(X_0 - a)^+]/(b-a).$$

Letting $N \to \infty$, we conclude that $\beta_\infty(a,b) = \lim_{N\to\infty} \beta_N(a,b)$ = the total number of downcrossings by the process $(X_0, X_1, ...)$, is finite a.s. Considering this for all rational $b > a$, we conclude that $\{X_n\}$ converges a.s., though possibly to $\pm\infty$. By conditional Jensen's inequality, $(X_n^+, \mathcal{B}_n), n \ge 0$, is also a reversed submartingale. Thus

$$E[\lim_{n\to\infty} X_n^+] \le \liminf_{n\to\infty} E[X_n^+] \le E[X_0^+] < \infty.$$

Hence $P(X_n \to \infty) = 0$. The first claim follows.

Let $X_\infty = \lim_{n\to\infty} X_n$ (possibly $-\infty$). If $\lim_{n\to\infty} E[X_n] = K > -\infty$,

$$E[\mid X_n \mid] = 2E[X_n^+] - E[X_n] \le 2E[X_0^+] - K.$$

Thus $\sup_n E[\mid X_n \mid] < \infty$ and by Fatou's lemma, $E[\mid X_\infty \mid] < \infty$. Let $\epsilon > 0$ and $m \ge 1$ be such that

$$E[X_n] < K + \epsilon, \; n \ge m.$$

Then for $n \ge m, a > 0$,

$$\begin{aligned}
\int_{\{|X_n|\geq a\}} | X_n | \, dP &= \int_{\{X_n\geq a\}} X_n \, dP + \int_{\{X_n\leq -a\}} X_n^- dP \\
&= \int_{\{X_n\geq a\}} X_n \, dP + \int_{\{X_n > -a\}} X_n dP - E[X_n] \\
&\leq \int_{\{X_n\geq a\}} X_m \, dP + \int_{\{X_n > -a\}} X_m dP - K \\
&= \int_{\{|X_n|\geq a\}} | X_m | \, dP + E[X_m] - K \\
&\leq \int_{\{|X_n|\geq a\}} | X_m | \, dP + \epsilon < 2\epsilon
\end{aligned}$$

for sufficiently large a, since $\sup_n P(| X_n | \geq a) \leq a^{-1} \sup_n E[| X_n |] \to 0$ as $a \to \infty$. Thus $\{X_n\}$ are u.i., implying $X_n \to X_\infty$ in L_1.

Finally, if $X_n \geq 0$ a.s. for all n and $E[(X_0)^p] < \infty$ for some $p \in (1, \infty)$, the arguments leading to Corollary 3.2.3 applied to the reversed sequence $X_n, ..., X_0$ lead to $E[\max_{0\leq n\leq N}(X_n)^p] \leq \left(\frac{p}{p-1}\right)^p E[(X_0)^p]$. Letting $N \to \infty$, one has $E[\sup_n (X_n)^p] < \infty$ and thus $E[(X_\infty)^p] < \infty$. Now using the dominated convergence theorem as in Corollary 3.3.1, we may conclude that $E[| X_n - X_\infty |^p] \to 0$ □

Corollary 3.3.3 *If L_1-convergence holds in the above, then $X_\infty \leq E[X_m/\mathcal{B}_\infty], m \geq 0$, where $\mathcal{B}_\infty = \cap_n \mathcal{B}_n$. In particular, if $(X_n, \mathcal{B}_n), n \geq 0$, is a reversed martingale, then $X_\infty = E[X_m/\mathcal{B}_\infty], m \geq 0$.*

Proof X_∞ is clearly $\mathcal{B}_\infty$-measurable. For $A \in \mathcal{B}_\infty$,

$$\int_A X_\infty dP = \lim_{n\to\infty} \int_A X_n dP \leq \int_A X_m dP, m \geq 0,$$

implying the first claim. The second claim follows similarly. □

Finally, we have the following result due to Blackwell and Dubins [6]:

Theorem 3.3.8 *Let $X_n, n = 1, 2, ..., \infty$ be real random variables on $(\Omega, \mathcal{F}, P)$ such that $X_n \to X_\infty$ a.s. and $E[\sup_n | X_n |] < \infty$. Let $\{\mathcal{F}_n\}$ be a family of sub-σ-fields of $\mathcal{F}$ which is either increasing or decreasing, with $\mathcal{F}_\infty = \vee_n \mathcal{F}_n$ or $\bigcap_n \mathcal{F}_n$ accordingly. Then $\lim_{n,j\to\infty} E[X_n/\mathcal{F}_j] = E[X_\infty/\mathcal{F}_\infty]$ a.s. and in L_1.*

Proof Let $Y_k = \sup_{n\geq k} X_n, k \geq 1$. Then for $n \geq k, X_n \leq Y_k$ and hence $E[X_n/\mathcal{F}_i] \leq E[Y_k/\mathcal{F}_i], i \geq 1$. Let

$$\begin{aligned}
Z_1 &= \lim_{j\to\infty} \sup_{i,n\geq j} E[X_n/\mathcal{F}_i], \\
Z_2 &= \lim_{j\to\infty} \inf_{i,n\geq j} E[X_n/\mathcal{F}_i].
\end{aligned}$$

Then by either the forward or reversed martingale convergence,

$$Z_1 \leq \lim_{j\to\infty} \sup_{i\geq j} E[Y_k/\mathcal{F}_i] = E[Y_k/\mathcal{F}_\infty], k \geq 1.$$

By the conditional dominated convergence theorem,

$$Z_1 \leq \lim_{k\to\infty} E[Y_k/\mathcal{F}_\infty] = E[X_\infty/\mathcal{F}_\infty].$$

Similarly, $Z_2 \geq E[X_\infty/\mathcal{F}_\infty]$, proving the a.s. convergence. L_1-convergence follows by the dominated convergence theorem. □

3.4 Martingale Inequalities

Another fertile arena in martingale theory is that of martingale inequalities. In addition to their uses in probability theory, they have implications in other fields such as harmonic analysis. We prove here what is perhaps the most celebrated specimen thereof, viz., the Burkholder inequalities, and state without proof an extension of them. We start with some notation and two technical lemmas.

Let $(\Omega, \mathcal{F}, P)$ be a probability space, $\{\mathcal{F}_n\}$ a filtration and $\mathcal{F}_\infty = \vee_n \mathcal{F}_n$. Let $\{X_n\}$ be real integrable random variables adapted to $\{\mathcal{F}_n\}$. Without loss of generality, we may take $X_0 = 0, \mathcal{F}_0 = \{\phi, \Omega\}$. Define also $X_n^* = \max_{0\leq j\leq n} \mid X_j \mid, X^* = \sup_n \mid X_n \mid, M_n = X_n - X_{n-1}, S_n = (\sum_{i=1}^n M_i^2)^{1/2}, n = 0, 1, 2, ..., \infty$. Let $\parallel Z \parallel_p = \sup_n E[\mid Z_n \mid^p]^{1/p}$ for any real-valued process $\{Z_n\}$.

Lemma 3.4.1 *Let* $\parallel X \parallel_1 < \infty$ *and* $(X_n, \mathcal{F}_n), n \geq 0$, *either a martingale or a nonnegative submartingale. Let* $T = min\{n \geq 1 \mid \mid X_n \mid \geq \lambda\}$ *for some* $\lambda > 0$ *(= ∞ when this set is empty). Then*

$$E[S_{T-1}^2] + E[X_{T-1}^2] < 2E[X_T X_{T-1}] \leq 2\lambda \parallel X \parallel_1 .$$

Proof By the optional sampling theorem,

$$E[\mid X_{T\wedge n} \mid] \leq E[\mid X_n \mid] \leq \lim_{m\to\infty} E[\mid X_m \mid].$$

By Fatou's lemma and the fact that $\mid X_{T-1} \mid < \lambda$,

$$E[\mid X_T X_{T-1} \mid] \leq \lambda E[\mid X_T \mid] \leq \lim_{n\to\infty} \lambda E[\mid X_n \mid] = \lambda \parallel X \parallel_1 .$$

Also,

$$S_{n-1}^2 + X_{n-1}^2 \quad = \quad 2 \sum_{1\leq j\leq k\leq n-1} M_j M_k$$

$$\begin{aligned}
&= 2\sum_{j=1}^{n-1} M_j(X_{n-1} - X_{j-1}) \\
&= 2X_{n-1}^2 - 2\sum_{j=1}^{n-1} X_{j-1}M_j \\
&= 2[X_n X_{n-1} - \sum_{j=1}^{n} X_{j-1}M_j].
\end{aligned}$$

Since $(\sum_{j=1}^{T\wedge n} X_{j-1}M_j, \mathcal{F}_{T\wedge n})$ is a martingale (Exercise 3.17),

$$E[S_{T\wedge n-1}^2 + X_{T\wedge n-1}^2] \leq 2E[X_{T\wedge n}X_{T\wedge n-1}].$$

But $\mid X_{T\wedge n}X_{T\wedge n-1} \mid \leq \lambda(\lambda + \mid X_T \mid)$ and $E[\mid X_T \mid] < \infty$. Thus dominated convergence theorem and Fatou's lemma allow us to let $n \to \infty$ in the above inequality to deduce the first inequality. □

Corollary 3.4.1 *In the above set-up,*

$$\begin{aligned}
\lambda P(S_\infty > \lambda, X^* \leq \lambda) &\leq 2 \parallel X \parallel_1, \\
\lambda P(S_\infty > \lambda) &\leq 3 \parallel X \parallel_1 .
\end{aligned}$$

Proof From Doob's maximal inequality,

$$\lambda P(X^* > \lambda) \leq \parallel X \parallel_1 .$$

Thus the second inequality follows from the first. To prove the latter, let T be as above. Then $S_{T-1} = S_\infty$ on the set $\{T = \infty\} = \{X^* \leq \lambda\}$ and by the above lemma,

$$\begin{aligned}
\lambda P(S_\infty > \lambda, X^* \leq \lambda) &\leq \lambda P(S_{T-1} > \lambda) \\
\leq \lambda^{-1}E[S_{T-1}^2] &\leq 2 \parallel X \parallel_1 .
\end{aligned}$$

□

Lemma 3.4.2 *Let $\{X_n\}$ above be a nonnegative submartingale. Let $\theta, \lambda > 0, p \in (1, \infty), \beta = \sqrt{1 + 2\theta^2}$ and $Y_n = (\theta S_n) \vee X_n^*, n \geq 1$. Then*

$$P(Y_n > \beta\lambda) \leq 3 \int_{\{Y_n > \lambda\}} X_n dP,$$

$$\parallel S_n \parallel_p \leq 9p^{3/2}(p-1)^{-1} \parallel X_n \parallel_p,$$

$$\parallel S_\infty \parallel_p \leq 9p^{3/2}(p-1)^{-1} \parallel X \parallel_p .$$

Proof Let $Z_n = I\{\theta S_n > \lambda\}, W_n = Z_n X_n, n \geq 1$. Then $Z_{n+1} \geq Z_n$ for all n and thus $(W_n, \mathcal{F}_n), n \geq 0$, is a nonnegative submartingale. Let $\sigma = min\{n \geq 1 \mid \theta S_n > \lambda\}(= \infty$ if the set is empty). On the set $\{\theta S_n > \beta\lambda, X_n^* \leq \lambda\}, \sigma \leq n, \max_{m\leq n} \mid W_m \mid\leq \lambda$ and

$$\mid M_\sigma \mid=\mid X_\sigma - X_{\sigma-1} \mid \leq \mid X_\sigma \mid \wedge \mid X_{\sigma-1} \mid\leq X_n^* \leq \lambda.$$

From the definition of β, the following is seen to hold on this set:

$$\begin{aligned}(1+2\theta^2)\lambda^2 &\leq \theta^2 S_n^2 = \theta^2 S_{\sigma-1}^2 + \theta^2 M_\sigma^2 + \theta^2 \sum_{j=\sigma+1}^{n} M_j^2 \\ &\leq \lambda^2 + \theta^2\lambda^2 + \theta^2 \sum_{j=\sigma+1}^{n} (W_j - W_{j-1})^2 \\ &\leq \lambda^2(1+\theta^2) + \theta^2 \sum_{j=1}^{n}(W_j - W_{j-1})^2.\end{aligned}$$

Thus $\sum_{j=1}^n (W_j - W_{j-1})^2 \geq \lambda^2$ on this set. Apply Corollary 3.4.1 to the nonnegative $\{\mathcal{F}_n\}$-submartingale $W_1, W_2, ..., W_n, W_n, ...,$ to obtain

$$\begin{aligned}\lambda P(Y_n > \beta\lambda, X_n^* \leq \lambda) &= \lambda P(\theta S_n > \beta\lambda, X_n^* \leq \lambda) \\ &\leq \lambda P(\sum_{j=1}^{n}(W_j - W_{j-1})^2 > \lambda^2, \max_{m\leq n} \mid W_m \mid\leq \lambda) \\ &\leq 2 \parallel W \parallel_1 = 2E[\mid W_n \mid] = 2E[\mid Z_n X_n \mid].\end{aligned}$$

As in Doob's maximal inequality,

$$\lambda P(X_n^* > \lambda) \leq \int_{\{X_n^* > \lambda\}} X_n dP \leq \int_{\{Y_n > \lambda\}} X_n dP.$$

Combining the estimates,

$$\begin{aligned}\lambda P(Y_n > \beta\lambda) &\leq \lambda P(X_n^* > \lambda) + \lambda P(\theta S_n > \beta\lambda, X_n^* \leq \lambda) \\ &\leq 3 \int_{\{Y_n > \lambda\}} X_n dP,\end{aligned}$$

which proves the first claim. Now

$$\begin{aligned}E[\mid Y_n \mid^p]/\beta^p &= p \int_0^\infty \lambda^{p-1} P(Y_n > \beta\lambda) d\lambda \\ &\leq 3p \int_0^\infty \lambda^{p-2} \int_{\{Y_n > \lambda\}} X_n dP d\lambda \\ &= 3pE[X_n \int_0^{Y_n} \lambda^{p-2} d\lambda] \\ &= \frac{3p}{p-1} E[X_n Y_n^{p-1}]\end{aligned}$$

$$\leq \frac{3p}{p-1} \parallel X_n \parallel_p \parallel Y_n \parallel_p^{p-1} .$$

Thus

$$\theta \parallel S_n \parallel_p \leq \parallel Y_n \parallel_p \leq 3p\beta^p(p-1)^{-1} \parallel X_n \parallel_p .$$

Let $\theta = 1/\sqrt{p}$. Then $\beta^p = (1+(2/p))^{p/2} < 3$. The second inequality follows. The third follows from the second by letting $n \to \infty$. □

Theorem 3.4.1 *(Burkholder) If $(X_n, \mathcal{F}_n), n \geq 0$, above is a martingale and $p \in (1, \infty)$, then for $A_p = [18p^{3/2}/(p-1)]^{-1}$ and $B_p = 18p^{3/2}/(p-1)^{1/2}$,*

$$A_p \parallel S_n \parallel_p \leq \parallel X_n \parallel_p \leq B_p \parallel S_n \parallel_p$$

$$A_p \parallel S_\infty \parallel_p \leq \parallel X \parallel_p \leq B_p \parallel S_\infty \parallel_p .$$

Proof The second set of inequalities follows from the first by letting $n \to \infty$, so we prove only the first. Fix n. Let $Y_j = E[X_n^+/\mathcal{F}_j], Z_j = E[X_n^-/\mathcal{F}_j]$ for $j \leq n$. Then $Y_n = X_n^+, Z_n = X_n^-$ and $X_j = E[X_n/\mathcal{F}_j] = Y_j - Z_j$ for $1 \leq j \leq n$. Let

$$S_n' = (\sum_{i=1}^{n}(Y_i - Y_{i-1})^2)^{1/2}, S_n'' = (\sum_{i=1}^{n}(Z_i - Z_{i-1})^2)^{1/2}, n \geq 1.$$

Then $S_n \leq S_n' + S_n''$ and by Minkowski's inequality and the above lemma,

$$\begin{aligned} \parallel S_n \parallel_p &\leq \parallel S_n' \parallel_p + \parallel S_n'' \parallel_p \\ &\leq 9p^{3/2}(p-1)^{-1}(\parallel Y_n \parallel_p + \parallel Z_n \parallel_p) \\ &\leq 18p^{3/2}(p-1)^{-1} \parallel X_n \parallel_p, \end{aligned}$$

which yields the first half of the first inequality.

To prove the second half, we may suppose without any loss of generality that $\parallel X_n \parallel_p > 0$ and $\parallel S_n \parallel_p < \infty$. Then $X_j \in L_p, 1 \leq j \leq n$. Let

$$\begin{aligned} W_n &= sgn(X_n) \mid X_n \mid^{p-1} / \parallel X_n \parallel_p^{p-1} \\ W_j &= E[W_n/\mathcal{F}_j], 1 \leq j \leq n. \end{aligned}$$

Then for $q = (1-1/p)^{-1}, W_j, 1 \leq j \leq n$, is an $\{\mathcal{F}_n\}$-martingale satisfying $\parallel W_j \parallel_q < \infty, 1 \leq j \leq n$. In fact, $\parallel W_n \parallel_q = 1$ and $E[W_n X_n] = \parallel X_n \parallel_p$. Let $V_1 = W_1, V_j = W_j - W_{j-1}$ for $2 \leq j \leq n$. Then for $n \geq 1$,

$$\begin{aligned} \parallel X_n \parallel_p &= E[X_n W_n] \\ &= E[(X_{n-1} + M_n)(W_{n-1} + V_n)] \\ &= E[X_{n-1}W_{n-1} + M_n V_n]. \end{aligned}$$

Iterating the argument and letting $\tilde{S}_n = (\sum_{j=1}^n (V_j - V_{j-1})^2)^{1/2}$,

$$\begin{aligned}
\| X_n \|_p &= E[\sum_{j=1}^{n} M_j V_j] \\
&\leq E[S_n \tilde{S}_n] \\
&\leq \| S_n \|_p \| \tilde{S}_n \|_q \\
&\leq \| S_n \|_p A_q^{-1} \| W_n \|_q \\
&\quad \text{(by the first half of the inequality already proved)} \\
&\leq \frac{18p^{3/2}}{(p-1)^{1/2}} \| S_n \|_p .
\end{aligned}$$

□

Corollary 3.4.2 *In the above set-up,*

$$A_p \| S_n \|_p \leq \| X_n^* \|_p \leq C_p \| S_n \|_p$$

$$A_p \| S_\infty \|_p \leq \| X^* \|_p \leq C_p \| S_\infty \|_p$$

with $C_p = 18p^{5/2}/(p-1)^{3/2}$.

Proof This is immediate from the above theorem and Corollary 3.2.3.□

We now state without proof the following generalization of the above theorem. A proof can be found in [8, Chapter 11].

Theorem 3.4.2 *(Burkholder–Davis–Gundy) Let* $\Phi : [0, \infty] \to [0, \infty]$ *be a nondecreasing function which is finite and convex on* $[0, \infty)$ *and satisfies* $\Phi(0) = 0, \Phi(\infty-) = \Phi(\infty)$ *and* $\Phi(2\lambda) \leq c\Phi(\lambda)$ *for all* $\lambda > 0$ *and some* $c > 0$. *Then there exist constants* $0 < A < B < \infty$ *depending on c only such that for any martingale* $(X_n, \mathcal{F}_n), n \geq 0$,

$$AE[\Phi(S_\infty)] \leq E[\Phi(X^*)] \leq BE[\Phi(S_\infty)].$$

In conclusion, it should be noted that despite their length, the two preceding sections show only a fraction of martingale-related convergence theorems and inequalities. The reader is urged to see [8, 12, 34] for more.

3.5 Additional Exercises

(3.18) Let X be a real random variable on $(\Omega, \mathcal{F}, P)$ with $E[X^2] < \infty$ and $\mathcal{G} \subset \mathcal{F}$ a sub-σ- field. Show that

$$E[(X - E[X/\mathcal{G}])^2/\mathcal{G}] \leq E[(X - Z)^2/\mathcal{G}]$$

for every $\mathcal{G}$-measurable real random variable Z satisfying $E[Z^2] < \infty$, and the equality holds if and only if $Z = E[X/\mathcal{G}]$ a.s.

(3.19) Let X be an integrable real random variable on $(\Omega, \mathcal{F}, P), \mathcal{G} \subset \mathcal{F}$ a sub-σ-field and Q another probability measure on $(\Omega, \mathcal{F})$ which is absolutely continuous with respect to P. Let $\Lambda = dQ/dP$ and $E[./\mathcal{G}], E_0[./\mathcal{G}]$ the conditional expectations under P, Q respectively. Show that $E_0[X/\mathcal{G}] = E[X\Lambda/\mathcal{G}]/E[\Lambda/\mathcal{G}]$.

(3.20) Let X, Y be integrable real random variables on $(\Omega, \mathcal{F}, P)$ such that $E[X/Y] = Y$ a.s., $E[Y/X] = X$ a.s.. Show that $X = Y$ a.s.

(3.21) Let $X_1, X_2, ...$ be real random variables defined on some probability space. Show that one can construct on some probability space random variables $\overline{X}_1, \overline{X}_2, ..., Z_1, Z_2, ...,$ such that $(X_1, X_2, ...)$ and $(\overline{X}_1, \overline{X}_2, ...)$ agree in law, $\{Z_i\}$ are i.i.d. uniformly distributed on [0,1] and $\overline{X}_{n+1} = f_n(\overline{X}_1, ..., \overline{X}_n, Z_n)$ for some measurable $f_n : \boldsymbol{R}^{n+1} \to \boldsymbol{R}$. (Hint: Use Corollary 3.1.2 and Exercise 1.19).

(3.22) Show that conditional independence need not be preserved under convergence in law (i.e., exhibit triplets of real random variables $(X_n, Y_n, Z_n), n = 1, 2, ..., \infty$, such that X_n, Y_n are conditionally independent given Z_n for $n = 1, 2, ...,$ but not so for $n = \infty$, and $(X_n, Y_n, Z_n) \to (X_\infty, Y_\infty, Z_\infty)$ in law.)

(3.23) Let $\{X_n\}$ be a sequence of Polish space valued random variables, $\{\mathcal{F}_n\}$ its natural filtration and T an $\{\mathcal{F}_n\}$-stopping time. Show that $\mathcal{F}_T = \sigma(X_{T\wedge n}, n \geq 0)$ modulo completion.

(3.24) For $i = 1, 2, .., \infty$, let $(M_0^i, M_1^i, ...)$ and $(Z_0^i, Z_1^i, ...)$ be respectively real and Polish space valued random variables on a probability space $(\Omega_i, \mathcal{F}_i, P_i)$ such that for $\mathcal{F}_n^i = \sigma(M_m^i, Z_m^i, m \leq n), n \geq 0, (M_n^i, \mathcal{F}_n^i)$, $n \geq 0$ is a martingale for $i = 1, 2, ...$ Furthermore, suppose that for each $n = 1, 2, ..., \{M_n^i, i \geq 0\}$ are u.i. and $(M_0^i, M_1^i, ..., Z_0^i, Z_1^i, ...) \to (M_0^\infty, M_1^\infty, ..., Z_0^\infty, Z_1^\infty, ...)$ in law as $i \to \infty$. Show that $(M_n^\infty, \mathcal{F}_n^\infty), n \geq 0$, is also a martingale.

(3.25) Let $(X_n, \mathcal{F}_n), n \geq 0$, be a nonnegative supermartingale. Show that $X_{n+i} = 0$ a.s. on $\{X_n = 0\}$ for all $i \geq 1$.

(3.26) Let $\{X_i\}$ be independent random variables on $(\Omega, \mathcal{F}, P)$ with $E[X_i] = 0, E[X_i^2] < \infty, \forall i$. Let $S_n = X_1 + ... + X_n, n \geq 1$. Prove "Kolmogorov's inequality"

$$P(\max_{1\leq m\leq n} | S_m | \geq \epsilon) \leq \epsilon^{-2} \sum_{n=1}^{n} E[X_i^2].$$

(3.27) Let $\{X_i\}$ be nonnegative integrable random variables on $(\Omega, \mathcal{F}, P)$ adapted to a filtration $\{\mathcal{F}_i\}$ and bounded a.s. by a constant $C < \infty$. Let $\mathcal{F}_{-1} = \{\phi, \Omega\}$. Show that the sets $\{\sum_{i=0}^{\infty} X_i < \infty\}$ and $\{\sum_{i=0}^{\infty} E[X_i/\mathcal{F}_{i-1}] < \infty\}$ coincide a.s.

(3.28) Let X, Y be real random variables on $(\Omega, \mathcal{F}, P)$ with $E[\| X \|] < \infty$. Let $Z_1, Z_2, ...$ be a sequence of proposed approximations of $E[X/Y]$ defined by

$$Z_n(\omega) = \sum_{k=-\infty}^{\infty} \frac{E\left[XI\left\{\frac{k}{2^n} \leq Y < \frac{k+1}{2^n}\right\}\right]}{P\left(\frac{k}{2^n} \leq Y < \frac{k+1}{2^n}\right)} I\left\{\frac{k}{2^n} \leq Y < \frac{k+1}{2^n}\right\}$$

where the ratio $E\left[XI\left\{\frac{k}{2^n} \leq Y < \frac{k+1}{2^n}\right\}\right]/P\left(\frac{k}{2^n} \leq Y < \frac{k+1}{2^n}\right)$ may be set equal to an arbitrary number when the denominator is zero. Show that $Z_n \to E[X/Y]$ a.s.

(3.29) Let $(X_n, \mathcal{F}_n), n \geq 0$, be a supermartingale and T an $\{\mathcal{F}_n\}$-stopping time a.s. bounded by $N < \infty$. Show that

$$E[\| X_T \|] \leq 3sup_{n \leq N} E[\| X_n \|].$$

(3.30) Let $(X_n, \mathcal{F}_n), n \geq 0$, be a uniformly integrable supermartingale. Show that $\{X_n\}$ can be written as $X_n = M_n + Z_n, n \geq 0$, where $\{M_n\}$ is a uniformly integrable $\{\mathcal{F}_n\}$-martingale and $\{Z_n\}$ is a uniformly integrable nonnegative $\{\mathcal{F}_n\}$-supermartingale satisfying $Z_n \to 0$ a.s. (This is called the Riesz decomposition.)

4

Basic Limit Theorems

4.1 Introduction

Limit theorems form a cornerstone of probability theory. These are results that describe the asymptotic behaviour of sequences of random variables, usually suitably normalized partial sums of another sequence of random variables. Even today a lot of research activity in the field is directed towards refining and extending them.

The classical triad of limit theorems consists of the law of large numbers which describes the "typical" or "average" behaviour of sums of random variables, the central limit theorem and its variants which describe the fluctuations around this typical behaviour, and the law of iterated logarithms which describes the upper and lower envelopes of these sums, all in a suitable asymptotic sense. The first and the third are a.s. convergence results while the second entails convergence in law. In addition, one has other related results like Cramer's theorem which dictates how rapidly the probabilities of certain untypical events decay. Finally, there are results like the three series theorem which serve as tests for convergence. Each of these subdomains is a vast field in itself. We shall only be seeing the tip of the iceberg when we treat below these results for sums of independent real-valued random variables. These have been extended in several directions — to dependent random variables (martingales, exchangeable random variables, functions of Markov chains etc.), abstract space-valued (e.g. Banach space-valued) random variables, "functional" limit theorems for stochastic processes, etc. These lie outside the scope of this book.

4.2 Strong Law of Large Numbers

Observe that square-integrable zero mean random variables on a probability space $(\Omega, \mathcal{F}, P)$ form a vector space with the inner product $\langle X, Y\rangle = E[XY]$. Adding n i.i.d. random variables from this collection is then tantamount to adding n mutually orthogonal vectors of the same length and thus the sum should be of the order $n^{1/2}$. Hence the arithmetic mean of such random variables should be of the order $n^{-1/2}$ and therefore go to zero as n $\to \infty$. The strong law of large numbers confirms this intuition. We shall study two variants of this law. The first concerns integrable, pairwise independent and identically distributed random variables $X_n, n \geq 1$, on a probability space. Let $S_n = \sum_{i=1}^n X_i, n \geq 1$.

Theorem 4.2.1 *(Strong law of large numbers)* $S_n/n \to E[X_1]$ *a.s.*

Proof (Etemadi) Since $\{X_n^+\}, \{X_n^-\}$ satisfy the assumptions of the theorem as well, we prove the result for $X_i \geq 0, i \geq 1$. Let $Y_i = X_i I\{X_i \leq i\}, i \geq 1$, and $S_n^* = \sum_{i=1}^n Y_i, n \geq 1$. Let $\epsilon > 0, \alpha > 1$ and $k_n = [\alpha^n](=$ the integer part of α^n) for $n \geq 1$. In what follows, C denotes a finite positive constant that can vary from step to step. Let $\mu =$ the law of X_i and $\sigma^2(...) =$ the variance of $\cdots$. Then

$$
\begin{aligned}
\sum_{n=1}^{\infty} P\left(\left|\frac{S_{k_n}^* - E[S_{k_n}^*]}{k_n}\right| > \epsilon\right) &\leq C\sum_{n=1}^{\infty}\frac{\sigma^2(S_{k_n}^*)}{k_n^2} \\
&= C\sum_{n=1}^{\infty}\frac{1}{k_n^2}\sum_{i=1}^{k_n}\sigma^2(Y_i) \\
&\leq C\sum_{i=1}^{\infty}\frac{E[Y_i^2]}{i^2} \\
&= C\sum_{i=1}^{\infty}\frac{1}{i^2}\int_0^i x^2\mu(dx) \\
&= C\sum_{i=1}^{\infty}\frac{1}{i^2}\sum_{k=0}^{i-1}\int_k^{k+1} x^2\mu(dx) \\
&\leq C\sum_{k=0}^{\infty}\frac{1}{k+1}\int_k^{k+1} x^2\mu(dx) \\
&\leq C\sum_{k=0}^{\infty}\int_k^{k+1} x\mu(dx) = CE[X_1] < \infty.
\end{aligned}
$$

Also,

$$E[X_1] = \lim_{n\to\infty} \int_0^n x\mu(dx) = \lim_{n\to\infty} E[Y_n] = \lim_{n\to\infty} \frac{E[S^*_{k_n}]}{k_n}.$$

Thus by the Borel–Cantelli lemma,

$$\frac{S^*_{k_n}}{k_n} \to E[X_1] \ a.s.$$

Now

$$\begin{aligned}
\sum_{n=1}^{\infty} P(Y_n \neq X_n) &= \sum_{n=1}^{\infty} P(X_n > n) \\
&= \sum_{n=1}^{\infty} \int_n^{\infty} \mu(dx) \\
&\leq \sum_{n=1}^{\infty} \sum_{i=n}^{\infty} \int_i^{i+1} \mu(dx) \\
&= \sum_{i=1}^{\infty} i \int_i^{i+1} \mu(dx) \\
&\leq E[X_1] < \infty.
\end{aligned}$$

By the Borel–Cantelli lemma, $P(X_n \neq Y_n \ i.o.) = 0$. Thus

$$\lim_{n\to\infty} \frac{S_{k_n}}{k_n} = E[X_1] \text{ a.s.}$$

For $n \geq 1$, let $m(n) \geq 0$ be such that $k_{m(n)} \leq n < k_{m(n)+1}$. Since $n \to S_n$ is nondecreasing,

$$\begin{aligned}
\liminf_{n\to\infty} \frac{S_n}{n} &\geq \liminf_{n\to\infty} \frac{S_{k_{m(n)}}}{k_{m(n)}} \frac{k_{m(n)}}{k_{m(n)+1}} \\
&\geq \frac{1}{\alpha} \lim_{n\to\infty} \frac{S_{k_{m(n)}}}{k_{m(n)}} = \frac{1}{\alpha} E[X_1] < \infty.
\end{aligned}$$

Similarly,

$$\limsup_{n\to\infty} \frac{S_n}{n} \leq \alpha E[X_1] \text{ a.s.}$$

Since $\alpha > 1$ was arbitrary, let $\alpha \to 1$ to conclude. □

In particular, this implies $S_n/n \to E[X_1]$ in probability, which is known as the weak law of large numbers. The second version of the strong law drops the requirement of identical distributions, but requires a uniform bound on $\{\sigma^2(X_i)\}$. Let $\{S_n\}$ be as before, $\tilde{S}_n = S_n - \sum_{i=1}^n E[X_i], i \geq 1$.

Theorem 4.2.2 *(Strong law of large numbers — second version) Let $\{X_i\}$ be independent with $\sup_i \sigma^2(X_i) < \infty$. Then $\tilde{S}_n/n \to 0$ a.s.*

This follows easily from Theorem 3.3.4 (Exercise 4.1). The next result shows that the integrability condition in Theorem 4.2.1 cannot be relaxed.

Theorem 4.2.3 *Let $\{X_i\}$ be i.i.d. (independent and identically distributed). Suppose $E[| X_1 |] = \infty$. Then S_n/n does not converge, a.s.*

Proof Suppose $P(S_n/n$ converges$) > 0$. Then by Kolmogorov's zero–one law, $P(S_n/n$ converges$) = 1$. But

$$\frac{X_n}{n} = \frac{S_n}{n} - \frac{n-1}{n}\frac{S_{n-1}}{n-1}.$$

Thus $X_n/n \to 0$ a.s., implying $P(| X_n |> n \; i.o.) = 0$. This is impossible by Exercise 1.25. □

A useful rephrasing of the law of large numbers is in terms of "empirical measures". Let $\{X_n\}$ be i.i.d. Polish space-valued random variables with common law $\mu \in \boldsymbol{P}(S), S$ being the Polish space in question. Define $\boldsymbol{P}(S)$-valued "empirical measures" $\{\mu_n, n \geq 1\}$ by

$$\mu_n(A) = \frac{1}{n}\sum_{m=1}^{n} I\{X_m \in A\}, \; A \text{ Borel in } S, n \geq 1.$$

Corollary 4.2.1 *$\mu_n \to \mu$ a.s.*

Proof Let $\{f_i\} \subset C_b(S)$ be a countable convergence determining class for $\boldsymbol{P}(S)$. Then the strong law implies

$$\int f_i d\mu_n = \frac{1}{n}\sum_{i=1}^{n} f_i(X_m) \to \int f_i d\mu \text{ a.s.}, i \geq 1.$$

The claim follows. □

For $S = \boldsymbol{R}$, we shall improve this to the following:

Theorem 4.2.4 *(Glivenko, Cantelli) $\mu_n((-\infty, x]) \to \mu((-\infty, x])$ a.s., uniformly in x.*

Proof Let $\epsilon > 0$. Let $\{x_i, 1 \leq i \leq n\} \subset \boldsymbol{R}, n \geq 1$, be such that $\mu(\{x_i\}) = 0$ for all $i, \mu((-\infty, x_1)) < \epsilon, \mu([x_i, x_{i+1})) < \epsilon$ for $1 \leq i < n$ and $\mu([x_n, \infty)) < \epsilon$. For each $x \in \boldsymbol{R}$, let $b(x) = x_i$ for the i for which $x_i \leq x < x_{i+1}$, where $x_0 = -\infty$ and $x_{n+1} = \infty$ by convention. Then $x \geq b(x)$ and

$$\begin{aligned}
&\liminf_{m\to\infty} \inf_x ((\mu_m((-\infty, x]) - \mu((-\infty, x])) \\
&\quad \geq \liminf_{m\to\infty} \inf_x (\mu_m((-\infty, b(x)]) - \mu((-\infty, b(x)])) \\
&\qquad + \inf_x (\mu((-\infty, b(x)]) - \mu((-\infty, x])).
\end{aligned}$$

Since $\mu_m((-\infty, x]) \to \mu((-\infty, x])$ for each x which satisfies $\mu(\{x\}) = 0$ by virtue of Corollary 4.2.1 above, it does so uniformly over $x \in \{x_1, ..., x_n\}$. Thus the first term on the right is zero. The second exceeds $-\epsilon$ by our choice of $b(x)$. Since $\epsilon > 0$ was arbitrary,

$$\liminf_{m\to\infty} \inf_x (\mu_m((-\infty, x]) - \mu((-\infty, x])) \geq 0.$$

A similar argument (using $b(x) = x_{i+1}$ when $x_i < x \leq x_{i+1}$) shows that

$$\limsup_{m\to\infty} \sup_x (\mu_m((-\infty, x]) - \mu((-\infty, x])) \leq 0.$$

Combine the two to conclude. □

For more general "uniform" convergence theorems along these lines, see [39].

4.3 Central Limit Theorem

Recall the intuition that the sum of n i.i.d. zero mean square-integrable random variables grows as $\sqrt{n}$. Thus if we divide it by $\sqrt{n}$, we would expect its law to remain in a tight set as n increases, possibly approaching a limit. The central limit theorem shows that this is indeed so. That the limit will be Gaussian (cf. Exercise 2.18) can be guessed from the following crude computation: Let $X_i, i \geq 0$, be i.i.d. with $E[X_i] = 0, E[X_i^2] = 1$ and let $S_n = \sum_{i=1}^n X_i, n \geq 1$. Let $\varphi_n(.)$ denote the characteristic function of $S_n/\sqrt{n}, n \geq 1$. Then

$$\begin{aligned} ln\varphi_n(\nu) &= lnE[e^{i\nu S_n/\sqrt{n}}] = nlnE[e^{i\nu X_1/\sqrt{n}}] \\ &\approx nlnE[1 + i\nu X_1/\sqrt{n} - \nu^2 X_1^2/2n + O(n^{-3/2})] \\ &= nln\left(1 - \frac{\nu^2}{2n} + O(n^{-3/2})\right) \\ &\approx -\nu^2/2 + O(n^{-1/2}) \end{aligned}$$

for large n. Thus $\varphi_n(\nu) \approx exp(-\nu^2/2)$ for large n, which is the characteristic function of $N(0,1)$. (This crude argument is quite close to the actual proof below.) Another "empirical" way to motivate the Gaussian limit is to take a density function $p(.)$ on $\boldsymbol{R}$ with zero mean and finite variance and compute $p*p, p*p*p$ and so on; where "$*$" denotes convolution. Then $p*p* \cdots * p$ (n times) is seen to approach a bell-shaped Gaussian curve for large n.

The theorem has a long history culminating in the version below due to Lindeberg and Feller. Consider a sequence of zero mean independent random variables $X_n, n \geq 1$, with variances $\sigma_n^2, n \geq 1$, respectively. (More generally, if $\{X_n\}$ are not zero mean, we replace X_n by $X_n - E[X_n]$ in everything that follows.) Let $S_n = \sum_{i=1}^n X_i, s_n^2 = E[S_n^2] = \sum_{i=1}^n \sigma_i^2$ for $n \geq 1$, with $s_n \geq 0$ (i.e., the positive root of s_n^2). Say that $\{X_n\}$ above obey the Lindeberg condition if $s_n^2 > 0$ for some n (and hence for all n thereafter) and for all $\epsilon > 0$,

$$\sum_{j=1}^n \int_{\{|X_j| > \epsilon s_j\}} X_j^2 dP = o(s_n^2). \tag{4.1}$$

Lemma 4.3.1 *The above condition is equivalent to*

$$\sum_{j=1}^n \int_{\{|X_j| > \epsilon s_n\}} X_j^2 dP = o(s_n^2). \tag{4.2}$$

Proof Both (4.1) and (4.2) imply $s_n \to \infty$. Since s_n increases with n, (4.1) implies (4.2) . Conversely, let (4.2) hold. For any $\epsilon > 0$ and $1 > \delta > 0$,

$$\begin{aligned}
&\frac{1}{s_n^2} \sum_{j=1}^n \int_{\{|X_j| > \epsilon s_j\}} X_j^2 dP \\
&= \frac{1}{s_n^2} \sum_{\{j \leq n | s_j \leq \delta s_n\}} \int_{\{|X_j| > \epsilon s_j\}} X_j^2 dP + \frac{1}{s_n^2} \sum_{\{j \leq n | s_j > \delta s_n\}} \int_{\{|X_j| > \epsilon s_j\}} X_j^2 dP \\
&\leq \delta^2 + \frac{1}{s_n^2} \sum_{j=1}^n \int_{\{|X_j| > \epsilon \delta s_n\}} X_j^2 dP \\
&\to \delta^2 \text{ as } s_n \to \infty.
\end{aligned}$$

Since δ was arbitrary, (4.1) holds. □

Lemma 4.3.2 *Under (4.1),* $\max_{1 \leq j \leq n} \frac{\sigma_j^2}{s_n^2} \to 0$ *as* $s_n \to \infty$.

Proof Let $\epsilon > 0$. Then

$$\begin{aligned}
\max_{1 \leq j \leq n} \sigma_j^2 / s_n^2 &\leq \max_{1 \leq j \leq n} \frac{1}{s_n^2} [\epsilon^2 s_n^2 + \int_{\{|X_j| > \epsilon s_n\}} | X_j |^2 dP] \\
&\to \epsilon^2 \text{ as } s_n^2 \to \infty.
\end{aligned}$$

The claim follows. □

Theorem 4.3.1 *(Central limit theorem) If (4.1) holds, $S_n/s_n \to N(0,1)$ in law and $\sigma_n/s_n \to 0$ as $s_n \to \infty$. Conversely, if $\sigma_n/s_n \to 0$ and $S_n/s_n \to N(0,1)$ in law, (4.1) must hold.*

Proof Suppose (4.1) holds. Then by Lemma 4.3.2, $\sigma_n/s_n \to 0$. Let $t \in \boldsymbol{R}, \epsilon > 0$ be fixed. Set

$$\begin{aligned} Y_j(t) &= exp(itX_j) - 1 - itX_j + t^2X_j^2/2, \\ a_j(t) &= exp(-\sigma_j^2t^2/2) - 1 + \sigma_j^2t^2/2. \end{aligned}$$

Now $\mid Y_j(t) \mid \leq \min(t^2X_j^2, \frac{1}{6} \mid tX_j \mid^3)$ (Exercise 4.2). Thus

$$\begin{aligned} &\left| E\left[e^{itX_j/s_n} - e^{-\sigma_j^2t^2/2s_n^2}\right]\right| \\ &= \left| E\left[Y_j\left(\frac{t}{s_n}\right) - a_j\left(\frac{t}{s_n}\right)\right]\right| \\ &\leq E\left[\left|\frac{tX_j}{s_n}\right|^2 I\{\mid X_j \mid > \epsilon s_n\} + \left|\frac{tX_j}{s_n}\right|^3 I\{\mid X_j \mid \leq \epsilon s_n\}\right] + \frac{\sigma_j^4t^4}{8s_n^4}, \end{aligned}$$

where we have used the inequality $e^{-a} - 1 + a \leq a^2/2$ for a ≥ 0. Set $S_0 = 0 = s_0$. Then for $1 \leq j \leq n$,

$$\begin{aligned} &\left| E\left[exp\left(it\left(\frac{S_j}{s_n}\right) + \frac{s_j^2t^2}{2s_n^2}\right)\right] - E\left[exp\left(it\left(\frac{S_{j-1}}{s_n}\right) + \frac{s_{j-1}^2t^2}{2s_n^2}\right)\right]\right| \\ &= \left| E\left[exp\left(it\left(\frac{S_{j-1}}{s_n}\right) + \frac{s_j^2t^2}{2s_n^2}\right)\right]\right| \left|E[exp(itX_j/s_n) - exp(-\sigma_j^2t^2/2s_n^2)]\right| \\ &\leq exp(t^2/2)\left(E\left[\frac{t^2X_j^2}{s_n^2}I\{\mid X_j \mid > \epsilon s_n\} + \frac{\epsilon \mid t \mid^3 X_j^2}{s_n^2}I\{\mid X_j \mid \leq \epsilon s_n\}\right]\right. \\ &\quad \left. + \frac{t^4\sigma_j^2}{s_n^2}\max_{1\leq j\leq n}\frac{\sigma_j^2}{s_n^2}\right). \end{aligned}$$

Hence

$$\begin{aligned} &\left| E\left[exp\left(it\frac{S_n}{s_n}\right) - exp(-t^2/2)\right]\right| \\ &\leq e^{-t^2/2}\left|\sum_{j=1}^{n}\left(E\left[exp\left(it\frac{S_j}{s_n} + \frac{s_j^2t^2}{2s_n^2}\right)\right]\right.\right. \\ &\quad \left.\left. - E\left[exp\left(it\frac{S_{j-1}}{s_n} + \frac{s_{j-1}^2t^2}{2s_n^2}\right)\right]\right)\right| \\ &\leq \frac{t^2}{s_n^2}\sum_{j=1}^{n} E[X_j^2 I\{\mid X_j \mid > \epsilon s_n\}] + \epsilon \mid t \mid^3 + t^4 \max_{1\leq j\leq n}\frac{\sigma_j^2}{s_n^2}. \end{aligned}$$

By Lemma 4.3.2 and (4.1), the right hand side tends to $\epsilon \mid t \mid^3$ as $s_n \to \infty$. Since $\epsilon > 0$ was arbitrary, the claim follows by Theorem 2.5.1.

For the converse, note that if $\sigma_n/s_n \to 0$ as $s_n \to \infty$, then

$$\max_{1\le j\le n} \frac{\sigma_j}{s_n} \le \max_{1\le j\le m} \frac{\sigma_j}{s_n} + \max_{m<j\le n} \frac{\sigma_j}{s_n} \to 0$$

as first n and then m tends to ∞. Let $\varphi_j(.)$ be the characteristic function of $X_j, j \ge 1$. Then it is seen that

$$\left|\varphi_j\left(\frac{t}{s_n}\right) - 1\right| \le t^2\sigma_j^2/2s_n^2$$

for $n \ge 1$ (Exercise 4.3), implying

$$\sum_{j=1}^n \left|\varphi_j\left(\frac{t}{s_n}\right) - 1\right| \le \sum_{j=1}^n \frac{t^2\sigma_j^2}{2s_n^2} = \frac{t^2}{2}.$$

Thus for large n,

$$\begin{aligned}\sum_{j=1}^n \left|ln\varphi_j\left(\frac{t}{s_n}\right) - \left(\varphi_j\left(\frac{t}{s_n}\right) - 1\right)\right| &\le \sum_{j=1}^n \left|\varphi_j\left(\frac{t}{s_n}\right) - 1\right|^2 \\ &\le \frac{t^4}{4} \max_{1\le j\le n} \frac{\sigma_j^2}{s_n^2} \to 0\end{aligned}$$

as $s_n \to \infty$. By hypothesis, $\frac{S_n}{s_n} \to N(0,1)$ in law. Thus $\Pi_{j=1}^n \varphi_j\left(\frac{t}{s_n}\right) \to exp\left(-\frac{t^2}{2}\right)$ pointwise by Theorem 2.5.1, leading to

$$\sum_{j=1}^n ln\varphi_j\left(\frac{t}{s_n}\right) \to -\frac{t^2}{2}$$

as $n \to \infty$. Hence

$$\sum_{j=1}^n \left(\varphi_j\left(\frac{t}{s_n}\right) - 1\right) + \frac{t^2}{2} \to 0$$

as $n \to \infty$. Taking real parts,

$$\begin{aligned}&\frac{t^2}{2} - \sum_{j=1}^n E\left[\left(1 - cos\left(\frac{tX_j}{s_n}\right)\right) I\{| X_j | \le \epsilon s_n\}\right] \\ &- \sum_{j=1}^n E\left[\left(1 - cos\left(\frac{tX_j}{s_n}\right)\right) I\{| X_j | > \epsilon s_n\}\right] \to 0 \qquad (4.3)\end{aligned}$$

as $n \to \infty$. But

$$\left|1 - cos\left(\frac{tX_j}{s_n}\right)\right| \le 2 \le \frac{2X_j^2}{\epsilon^2 s_n^2} \text{ on } \{| X_j | > \epsilon s_n\}$$

and

$$\left|1 - \cos\left(\frac{tX_j}{s_n}\right)\right| \leq \frac{t^2 X_j^2}{2s_n^2}$$

(Exercise 4.4). Thus

$$\sum_{j=1}^{n} E\left[\left(1 - \cos\left(\frac{tX_j}{s_n}\right)\right) I\{| X_j |> \epsilon s_n\}\right] \leq \sum_{j=1}^{n} \frac{2\sigma_j^2}{\epsilon^2 s_n^2} = \frac{2}{\epsilon^2},$$

$$\sum_{j=1}^{n} E\left[\left(1 - \cos\left(\frac{tX_j}{s_n}\right)\right) I\{| X_j |\leq \epsilon s_n\}\right] \leq \frac{t^2}{2s_n^2} \sum_{j=1}^{n} E\left[X_j^2 I\{| X_j |\leq \epsilon s_n\}\right].$$

From (4.3), we then have

$$\frac{t^2}{2}\left(1 - \frac{1}{s_n^2}\sum_{j=1}^{n} E\left[X_j^2 I\{| X_j |\leq \epsilon s_n\}\right]\right) \leq \frac{2}{\epsilon^2} + o(1).$$

Therefore

$$\liminf_{n\to\infty} \frac{1}{s_n^2}\sum_{j=1}^{n}\int_{\{|X_j|\leq \epsilon s_n\}} X_j^2 dP \geq 1 - \frac{4}{t^2\epsilon^2} \to 1$$

as $t \to \infty$. Thus (4.2) and therefore (4.1) follows. □

The following theorem due to Berry and Essen gives an estimate of the rate of convergence in the central limit theorem.

Theorem 4.3.2 *In the foregoing, suppose that for some $\delta > 0$,*

$$\xi_n^{2+\delta} = \sum_{i=1}^{n} E\left[| X_i |^{2+\delta}\right] < \infty, \quad n \geq 1.$$

Then there exists a universal constant C_δ such that

$$\sup_{x\in R}\left|P\left(\frac{S_n}{s_n} < x\right) - \frac{1}{\sqrt{2\pi}}\int_{-\infty}^{x} e^{-x^2/2} dy\right| \leq C_\delta \left(\frac{\xi_n}{s_n}\right)^{2+\delta}.$$

See [8, Ch. 9] for a proof. The theorem is useful when $\frac{\xi_n}{s_n} \to 0$ as $n \to \infty$.

4.4 The Law of Iterated Logarithms

The law of iterated logarithms describes how the upper and lower envelopes of sums of independent random variables grow asymptotically. It has its origins in certain number theoretic studies of Borel and others and is a truly spectacular achievement of probability theory, particularly since there are no simple intuitive pointers to such a law. It can be proved under various sets of conditions, some of them nonoverlapping. We shall use one particular set of conditions, following [9]. The set-up is the same as in the preceding section, with $s_n \to \infty$ as $n \to \infty$.

In what follows, $C > 0$ will denote a constant that can vary from step to step. Let $\gamma_n = E[| X_n |^3], \Gamma_n = \sum_{i=1}^{n} \gamma_i, n \geq 1$, and $\varphi(\lambda, x) = \sqrt{2\lambda x^2 lnlnx}$, $\lambda, x > 0$.

Lemma 4.4.1 *Suppose that for some ϵ in $(0, 1)$,*

$$\frac{\Gamma_n}{s_n^3} \leq \frac{C}{(lns_n)^{1+\epsilon}}. \tag{4.4}$$

Then for each $\delta \in (0, \epsilon)$,

$$P(S_n > \varphi(1+\delta, s_n)) \leq C/(lns_n)^{1+\delta},$$
$$P(S_n > \varphi(1-\delta, s_n)) \geq C/(lns_n)^{1-(\delta/2)}.$$

Proof By Theorem 4.3.2,

$$P(S_n > xs_n) \approx \frac{1}{\sqrt{2\pi}} \int_x^\infty e^{-y^2/2} dy + C\frac{\Gamma_n}{s_n^3} \tag{4.5}$$

for all x. Since for $x > 0$,

$$\frac{x}{1+x^2} e^{-x^2/2} \leq \int_x^\infty e^{-y^2/2} dy \leq \frac{1}{x} e^{-x^2/2}$$

(Exercise 4.5), one has $\int_x^\infty e^{-y^2/2} dy \approx e^{-x^2/2}/x$ as $x \to \infty$. Let $x = \sqrt{2(1 \pm \delta) lnlns_n}$. Then

$$\frac{1}{\sqrt{2\pi}} \int_x^\infty e^{-y^2/2} dy \approx \frac{1}{\sqrt{4\pi(1 \pm \delta) lnlns_n}} \frac{1}{(lns_n)^{1\pm\delta}}. \tag{4.6}$$

Since $0 < \delta < \epsilon$ and

$$\frac{\Gamma_n}{s_n^3} \leq \frac{C}{(lns_n)^{1+\epsilon}}, \tag{4.7}$$

we have, from (4.5), (4.6), (4.7),

$$P(S_n > \sqrt{2(1 \pm \delta) lnlns_n}\; s_n) \approx \frac{C}{\sqrt{4\pi(1 \pm \delta) lnlns_n}} \frac{1}{(lns_n)^{1\pm\delta}}.$$

The claim follows. □

Before we proceed, we need the following technical lemma:

Lemma 4.4.2 *Let $\{E_i\}, \{F_i\}, 1 \leq i \leq n < \infty$, be events such that for each j, F_j is independent of $E_1^c E_2^c \ldots E_{j-1}^c E_j$ and there exists a constant $a > 0$ such that $P(F_j) \geq a$ for all j. Then*

$$P\left(\bigcup_{j=1}^{n} E_j F_j\right) \geq aP\left(\bigcup_{j=1}^{n} E_j\right).$$

Proof We have

$$\begin{aligned} P\left(\bigcup_{j=1}^{n} E_j F_j\right) &= P\left(\bigcup_{j=1}^{n} [(E_1F_1)^c \ldots (E_{j-1}F_{j-1})^c (E_jF_j)]\right) \\ &\geq P\left(\bigcup_{j=1}^{n} [E_1^c \ldots E_{j-1}^c E_j F_j]\right) \\ &= \sum_{j=1}^{n} P(E_1^c \ldots E_{j-1}^c E_j) P(F_j) \\ &\geq a \sum_{j=1}^{n} P(E_1^c \ldots E_{j-1}^c E_j) \\ &= aP\left(\bigcup_{j=1}^{n} E_j\right), \end{aligned}$$

which is the desired result. □

Let $E_n^+ = \{\omega \mid S_n(\omega) > \varphi(1+\delta, s_n)\}$ and $E_n^- = \{\omega \mid S_n(\omega) > \varphi(1-\delta, s_n)\}$, where "$\omega$" is a typical sample point of the underlying probability space.

Lemma 4.4.3 $P(E_n^+ \text{ } i.o.) = 0$.

Proof Let $c > 1$ and $\{n_k\} \subset \{n\}$ such that $s_{n_k} \leq c^k < s_{n_k} + 1$ for each k. Condition (4.4) above implies Lindberg's condition (Exercise 4.6) and thus $\max_{1 \leq j \leq n} \frac{\sigma_j}{s_n} \to 0$ by Lemma 4.3.2. Thus $s_{n_k+1}/s_{n_k} \to 1$. Hence $s_{n_k} \sim c^k$ as $k \to \infty$. Let $k \geq 1$, $F_j = \{\omega : | S_{n_{k+1}}(\omega) - S_j(\omega) | < s_{n_{k+1}}\}, n_k \leq j < n_{k+1}$. By Chebyshev's inequality,

$$P(F_j) \geq 1 - \frac{s_{n_{k+1}}^2 - s_{n_k}^2}{s_{n_{k+1}}^2} \to \frac{1}{c^2}$$

as $k \to \infty$. Hence $P(F_j) \geq \frac{1}{2c^2}$ for sufficiently large k, j being as above. By the preceding lemma,

$$P\left(\bigcup_{j=n_k}^{n_{k+1}-1} E_j^+ F_j\right) \geq \frac{1}{2c^2} P\left(\bigcup_{j=n_k}^{n_{k+1}-1} E_j^+\right)$$

for large k. Now

$$E_j^+ \bigcap F_j \subset \{\omega \mid S_{n_{k+1}}(\omega) > S_j(\omega) - s_{n_{k+1}} > \varphi(1+\delta, s_j) - s_{n_{k+1}}\}.$$

For $\beta \in (0,1)$ and sufficiently large k,

$$\varphi(1+\delta, s_j) - s_{n_{k+1}} > \varphi\left(\beta\left(\frac{1+\delta}{c^2}\right), s_{n_{k+1}}\right).$$

for j between n_k and $n_{k+1} - 1$ (Exercise 4.7). Pick β sufficiently close to 1 and $c > 1$ such that

$$\beta\left(\frac{1+\delta}{c^2}\right) > 1 + \frac{\delta}{2}$$

and set

$$G_k = \{\omega \mid S_{n_{k+1}}(\omega) > \varphi(1 + \frac{\delta}{2}, s_{n_{k+1}})\}.$$

Then G_k is same as $E_{n_{k+1}}^+$ with $\frac{\delta}{2}$ replacing δ. From the foregoing, $E_j^+ F_j \subset G_k$ for sufficiently large k, with j being between n_k and $n_{k+1} - 1$ as before. Thus

$$\bigcup_{j=n_k}^{n_{k+1}-1} E_j^+ F_j \subset G_k.$$

By Lemma 4.4.1,

$$\begin{aligned}
\sum_{k=1}^{\infty} P(G_k) &\leq \sum_{k=1}^{\infty} \frac{C}{(ln s_{n_k})^{1+(\delta/2)}} \\
&\leq C \sum_{k=1}^{\infty} (klnc)^{-(1+\delta/2)} < \infty.
\end{aligned}$$

Thus

$$\begin{aligned}
\sum_{k=1}^{\infty} P\left(\bigcup_{j=n_k}^{n_{k+1}-1} E_j^+\right) &\leq C \sum_{k=1}^{\infty} P\left(\bigcup_{j=n_k}^{n_{k+1}-1} E_j^+ Fj\right) \\
&\leq \sum_{k=1}^{\infty} P(G_k) < \infty.
\end{aligned}$$

By the Borel–Cantelli lemma, $P(\bigcup_{j=n_k}^{n_{k+1}-1} E_j^+ \text{ i.o.}) = 0$. This is equivalent to the claim (Exercise 4.8). □

Let $\{n_k\}$ be as above, but for an arbitrary $c > 1$. Set $t_k^2 = s_{n_{k+1}}^2 - s_{n_k}^2$ and

$$D_k = \left\{\omega \mid S_{n_{k+1}}(\omega) - S_{n_k}(\omega) > \varphi\left(1 - \frac{\delta}{2}, t_k\right)\right\}.$$

Lemma 4.4.4 $P(D_k\ i.o.) = 1.$

Proof Note that $S_{n_{k+1}} - S_{n_k}, k \geq 1$, are independent and hence so are $\{D_k\}$. Also,

$$t_k^2 \sim \left(1 - \frac{1}{c^2}\right) s_{n_{k+1}}^2 \sim \left(1 - \frac{1}{c^2}\right) c^{2(k+1)}.$$

Hence by (4.4),

$$\frac{\Gamma_{n_{k+1}} - \Gamma_{n_k}}{t_k^3} \leq \frac{C\Gamma_{n_{k+1}}}{s_{n_{k+1}}^3} \leq \frac{C}{(lnt_k)^{1+\epsilon}}.$$

By Lemma 4.4.1 applied to $S_{n_k}, S_{n_k+1}, \ldots,$

$$P(D_k) \geq \frac{C}{(lnt_k)^{1-\delta/4}} \geq \frac{C}{k^{1-\delta/4}}.$$

Thus $\sum_{k=1}^{\infty} P(D_k) = \infty$. The claim now follows from the Borel–Cantelli lemma. □

Lemma 4.4.5 $P(E_n^-\ i.o.) = 1.$

Proof From Lemma 4.4.3 applied to $\{-X_n\}$ with δ =1 and Lemma 4.4.4, the following hold a.s.:

$$\begin{aligned} S_{n_{k+1}} - S_{n_k} &> \varphi\left(1 - \frac{\delta}{2}, t_k\right) \text{ i.o.,} \\ S_{n_k} &\geq -\varphi(2, s_{n_k}) \end{aligned}$$

for k sufficiently large. Thus almost surely,

$$S_{n_{k+1}} > \varphi\left(1 - \frac{\delta}{2}, t_k\right) - \varphi(2, s_{n_k})\ i.o.$$

Since $lnlnt_k^2 \sim lnlns_{n_k}^2$ and $s_{n_k} \sim c^k$, the right hand side asymptotically exceeds

$$\alpha\left(\sqrt{\left(1 - \frac{\delta}{2}\right)\left(1 - \frac{1}{c^2}\right)} - \sqrt{\frac{2}{c^2}}\right)\varphi(1, s_{n_{k+1}}) > \varphi(1 - \delta, s_{n_{k+1}}),$$

for $0 < \alpha < 1$, provided c is taken sufficient large and α sufficiently close to 1. Thus $P(E_{n_{k+1}}^-\ \text{i.o.}) = 1$, which implies the claim. □

Theorem 4.4.1 *(Law of iterated logarithms) Under (4.4),*

$$\limsup_{n\to\infty} \frac{S_n}{\sqrt{2s_n^2 lnlns_n}} = 1 \; a.s.,$$
$$\liminf_{n\to\infty} \frac{S_n}{\sqrt{2s_n^2 lnlns_n}} = -1 \; a.s..$$

Proof By Lemmas 4.4.3 and 4.4.5,

$$1-\delta \leq \limsup_{n\to\infty} \frac{S_n}{\sqrt{2s_n^2 lnlns_n}} \leq 1+\delta \text{ a.s.}$$

for $\delta < \epsilon$. The first claim follows by considering $\delta = \epsilon/n, n = 2, 3, \ldots$. The second follows from the first by replacing $\{X_n\}$ by $\{-X_n\}$. □

See [4, 28] for further ramifications of the law of iterated logarithms. We mention in passing that for the law to hold for i.i.d. zero mean random variables, it is both necessary and sufficient that they be square-integrable [44].

4.5 Large Deviations

Large deviations refers to that class of results which describe how the probabilities of untypical events (or of "large deviations" away from the typical events) decay to zero. An abstract statement of the large deviations principle is as follows: Given a Polish space S and $\{\mu_\epsilon, \epsilon > 0\} \subset \boldsymbol{P}(S)$, the family $\{\mu_\epsilon, \epsilon > 0\}$ is said to satisfy a large deviations principle with rate function $I : S \to [0, \infty]$ if for B Borel in S,

$$-\inf_{B^\circ} I \leq \liminf_{\epsilon\to 0} \epsilon ln\mu_\epsilon(B) \leq \limsup_{\epsilon\to 0} \epsilon ln\mu_\epsilon(B) \leq -\inf_{\overline{B}} I$$

where B° (resp. $\overline{B}$) is the interior (resp. closure) of B and the infimum over an empty set is taken to be ∞ by convention. To see what this entails, let B be a small neighborhood of $x \in S$ and suppose that $\inf_{B^\circ} I = \inf_{\overline{B}} I = \alpha > 0$. Then $\mu_\epsilon(B) \sim exp(-\alpha/\epsilon)$, which dictates how the probability mass near x decays as $\epsilon \to 0$.

One of the most important instances of large deviations is Cramer's theorem, which is the large deviations principle associated with the strong law of large numbers. That is, for $\{X_i\}$ i.i.d. real valued zero mean random variables and $S_n = \sum_{i=1}^n X_i, n \geq 1$, it dictates how $P\left(\frac{S_n}{n} \in A\right)$ decays with n when $0 \notin A$. Our treatment thereof follows [13]. Before we state and prove this result, we introduce the candidate "rate function" $I : \boldsymbol{R} \to \boldsymbol{R}\bigcup\{\infty\}$ defined by

$$I(x) = \sup_{\theta}(\theta x - lnE[exp(\theta X)]) \text{ (possibly } \infty).$$

One checks that $J : \theta \to \ln E[\exp(\theta X)] \in [0, \infty]$ is convex lower semicontinuous and hence so is $I(.)$ (Exercise 4.9), which is its "Legendre transform" in convex analysis parlance. Also, $I(.) \geq 0$ (Exercise 4.10). The following lemma lists some additional properties of $I(.)$.

Lemma 4.5.1 *(i) If $E[\| X \|] < \infty$ and $m = E[X]$, then $I(m) = 0, I : [m, \infty) \to [0, \infty]$ is nondecreasing and $I : (-\infty, m] \to [0, \infty]$ is nonincreasing.*

(ii) For $x \geq m, I(x) = \sup_{\theta \geq 0}(\theta x - J(\theta))$ and $P(X \geq x) \leq exp(-I(x))$. For $x \leq m, I(x) = \sup_{\theta \leq 0}(\theta x - J(\theta))$ and $P(X \leq x) \leq exp(-I(x))$.

(iii) If $J(x) < \infty$ for all x in a neighborhood of 0, then $I(x) \to \infty$ as $| x | \to \infty$.

(iv) If $J(x) < \infty$ for all x, then $J(.) \in C^{\infty}(\boldsymbol{R})$ and $I(x)/ | x | \to \infty$ as $| x | \to \infty$.

Proof (i) By Jensen's inequality, $E[\exp(\theta X)] \geq \exp(\theta E[X])$. Thus $J(\theta) \geq \theta m$ for all θ, implying $\theta m - J(\theta) \leq 0$ for all θ. Thus $I(m) \leq 0$. Since $I(.) \geq 0$ anyway, $I(m) = 0$. Also, the facts that $I(.)$ is convex nonnegative and $I(m) = 0$ together imply that $I(.)$ is nonincreasing on $(-\infty, m]$ and nondecreasing on $[m, \infty)$.
(ii) Since $J(\theta) \geq \theta m$ and $I(.) \geq 0$, $I(x) = \sup_{\theta}(\theta x - J(\theta)) = \sup_{\theta \geq 0}(\theta x - J(\theta))$ for $x \geq m$. Similarly, $I(x) = \sup_{\theta \leq 0}(\theta x - J(\theta))$ for $x \leq m$. By Chebyshev's inequality, for $x \geq m$ and $\theta \geq 0$, we have

$$P(X \geq x) \leq E[exp(\theta X)]/exp(\theta x) = exp(-(\theta x - J(\theta)).$$

Hence $P(X \geq x) \leq exp(-I(x))$. Similarly for $x \leq m, P(X \leq x) \leq exp(-I(x))$.
(iii) Let $\overline{\theta} > 0$ and $J(\overline{\theta}) < \infty$. Then

$$\begin{aligned} \liminf_{x \to \infty} \frac{I(x)}{x} &= \liminf_{x \to \infty} \sup_{\theta}(\theta x - J(\theta))/x \\ &\geq \liminf_{x \to \infty} \frac{\overline{\theta} x - J(\overline{\theta})}{x} = \overline{\theta}. \end{aligned}$$

Thus $\lim_{x \to \infty} I(x) = \infty$. For $\overline{\theta} < 0$ and $J(\overline{\theta}) < \infty$, an analogous argument leads to $\limsup_{x \to -\infty} I(x)/x \leq \overline{\theta}$ and thus $\lim_{x \to -\infty} I(x) = \infty$.
(iv) If $J(\theta) < \infty$ for all θ, the claim: $I(x)/ \mid x \mid \to \infty$ as $\mid x \mid \to \infty$, follows as above. If $J(\pm\overline{\theta}) < \infty$ for some $\overline{\theta} < \infty$, one can easily verify using

the dominated convergence theorem that for $\theta \in (-\overline{\theta}, \overline{\theta}), \lim_{\Delta \to \infty}((J(\theta + \Delta) - J(\theta))/\Delta) = \frac{dJ(\theta)}{d\theta}$ exists and equals $E[X\exp(\theta X)]/exp(J(\theta))$ (Exercise 4.11). Repeating the argument, one sees that $J(.) \in C^\infty(\boldsymbol{R})$. □

Let $\{X_i\}$ be i.i.d. with law ν, $S_n = \sum_{i=1}^n X_i, n \geq 1$ and let $\mu_n =$ the law of S_n/n for $n \geq 1$. Let $J(\theta) = \ln E[\exp(\theta X_1)], \theta \in \boldsymbol{R}$. Define $I(.)$ as above.

Lemma 4.5.2 *If* $E[\| X_1 \|] < \infty$, *then* $\limsup_{n\to\infty} \frac{1}{n} ln(\mu_n(F)) \leq -\inf_F I(.)$ *for all closed* $F \subset \boldsymbol{R}$.

Proof Let $m = E[X_1] = E[S_n/n] = m, n \geq 1$. If $J_n(\theta) = \ln E[\exp(\theta S_n/n)]$, then $J_n(.) = nJ(./n), n \geq 1$ (Exercise 4.12). Hence $I_n(x) = \sup_\theta(\theta x - J_n(\theta)) = nI(x), x \in \boldsymbol{R}$. Let $x \geq m$ (resp. $y \leq m$). Then by (ii) above,

$$P(S_n/n \geq x) \leq exp(-nI(x))$$

(resp. $P(S_n/n \leq y) \leq exp(-nI(y))$). Recall that $I(.)$ is nondecreasing on $[m, \infty)$ and nonincreasing on $(-\infty, m]$. Thus the claim is immediate when $F \subset$ either $(-\infty, m]$ or $[m, \infty)$. If not, let $a = \inf\{x \geq m \mid x \in F\}, b = sup\{x \leq m \mid x \in F\}$. Then

$$P\left(\frac{S_n}{n} \in F\right) \leq exp(-nI(b)) + exp(-nI(a)) \leq 2exp(-n(\inf_F I)).$$

The claim follows. □

Theorem 4.5.1 *(Cramer) Suppose* $I(\theta) < \infty, \theta \in \boldsymbol{R}$. *Then for all Borel* $B \subset \boldsymbol{R}$,

$$-\inf_{B^\circ} I \leq \liminf_{n\to\infty} \frac{1}{n} ln(\mu_n(B)) \leq \limsup_{n\to\infty} \frac{1}{n} ln(\mu_n(B)) \leq -\inf_{\overline{B}} I$$

where the infimum over an empty set is infinity by convention.

Proof We shall first prove that if $\overline{x} \in \boldsymbol{R}$ and $\delta > 0$, then

$$liminf_{n\to\infty} \frac{1}{n} lnP\left(\left|\frac{S_n}{n} - \overline{x}\right| < \delta\right) \geq -I(\overline{x}).$$

If this holds, it follows that for any open set G,

$$liminf_{n\to\infty} \frac{1}{n} lnP\left(\frac{S_n}{n} \in G\right) \geq -\inf_G I.$$

To prove the former, first suppose that there is a $\overline{\theta} \in \boldsymbol{R}$ such that $I(\overline{x}) = \overline{\theta}\overline{x} - J(\overline{\theta})$. Consider the probability measure

$$\tilde{\nu}(dx) = [exp(\overline{\theta}x)/exp(J(\overline{\theta}))]\nu(dx).$$

If $\{X'_n\}$ were i.i.d. with law $\tilde{\nu}$, one sees that the law of $(X'_1 + \cdots + X'_n)/n$ is given (Exercise 4.13) by

$$\tilde{\mu}_n(dx) = [exp(n\overline{\theta}x)/exp(J_n(n\overline{\theta}))]\mu_n(dx), n \geq 1.$$

Then $\int | x | \tilde{\nu}(dx) < \infty$ (Exercise 4.14) and

$$\int x d\tilde{\nu}(x) = \int xe^{\overline{\theta}x}\nu(dx)/\int e^{\overline{\theta}x}\nu(dx) = \frac{d}{dt}(J(t))|_{t=\overline{\theta}}.$$

Since the map $t \to \overline{x}t - J(t)$ achieves its maximum at $t = \overline{\theta}$ by hypothesis, $\frac{d}{dt}(\overline{x}t - J(t)) = 0$ at $t = \overline{\theta}$. Thus $\overline{x} = \int x d\tilde{\nu}(x)$. By the weak law of large numbers,

$$\tilde{\mu}_n((\overline{x} - \delta, \overline{x} + \delta)) \to 1 \text{ as } n \to \infty.$$

If $\overline{\theta} \geq 0$, we have for $n \geq 1$,

$$\begin{aligned} P\left(\left| \frac{S_n}{n} - \overline{x} \right| < \delta\right) &= \int_{\{|\frac{S_n}{n} - \overline{x}| < \delta\}} dP \\ &\geq \int_{\{|\frac{S_n}{n} - \overline{x}| < \delta\}} exp(\overline{\theta}(S_n - n(\overline{x} + \delta)))dP \\ &= e^{-n(\overline{\theta}(\overline{x}+\delta) - J(\overline{\theta}))}\tilde{\mu}_n\left((\overline{x} - \delta, \overline{x} + \delta)\right) \\ &\geq e^{-n\overline{\theta}\delta - nI(\overline{x})}\tilde{\mu}_n\left((\overline{x} - \delta, \overline{x} + \delta)\right). \end{aligned}$$

Thus

$$liminf_{n\to\infty}\frac{1}{n}lnP\left(\left|\frac{S_n}{n} - \overline{x}\right| < \delta\right) \geq -I(\overline{x}) - \overline{\theta}\delta.$$

Since $\delta > 0$ was arbitrary and the left hand side is nondecreasing in δ, we can let $\delta \to 0$ on the right hand side only to conclude. If there is a $\overline{\theta} \leq 0$ such that $I(\overline{\theta}) = \overline{\theta}\overline{x} - J(\overline{\theta})$, a similar proof works with $\overline{x} - \delta$ replacing $\overline{x} + \delta$. Now suppose that $I(\overline{x}) > \theta\overline{x} - J(\theta)$ for all θ. If $\overline{x} \geq m$, then since $I(x) = \sup_{\theta\geq 0}(\theta x - J(\theta))$ on $[m, \infty)$, we must have $\theta_n \to \infty$ such that $\theta_n\overline{x} - J(\theta_n) \to I(\overline{x})$. Then

$$E[\exp(\theta_n(X_1 - \overline{x}))I\{X_1 < \overline{x}\}] \to 0.$$

Also,

$$E[\exp(-\theta_n\overline{x} + \theta_n X_1)] = \exp(-(\theta_n\overline{x} - J(\theta_n))) \to \exp(-I(\overline{x})).$$

Together, these lead to

$$E[\exp(\theta_n(X_1 - \overline{x}))I\{X_1 \geq \overline{x}\}] \to \exp(-I(\overline{x})).$$

But this is possible only if $P(X_1 > \overline{x}) = 0$ and $P(X_1 = \overline{x}) = \exp(-I(\overline{x}))$. Hence $P\left(\frac{S_n}{n} = \overline{x}\right) \geq \exp(-nI(\overline{x}))$ and

$$\liminf_{n\to\infty} \frac{1}{n} lnP\left(\left|\frac{S_n}{n} - \overline{x}\right| < \delta\right) \geq \liminf_{n\to\infty} \frac{1}{n} lnP\left(\frac{S_n}{n} = \overline{x}\right) \geq -I(\overline{x}).$$

Finally, a similar argument holds for $\overline{x} < m$, proving the claim made at the beginning of this proof. This claim in conjunction with the preceding lemma completes the proof of the theorem. □

As already remarked, Cramer's theorem is a large deviations principle that goes with the strong law of large numbers. One similarly has a large deviations principle to go with Corollary 4.2.1. Let $\{X_n\}$ be i.i.d. random variables taking values in a Polish space S. Let μ be their common law. Define the $\boldsymbol{P}(S)$-valued process of empirical measures $\{\mu_n\}$ as in Corollary 4.2.1. Let $\eta_n \in \boldsymbol{P}(\boldsymbol{P}(S))$ be the law of μ_n for $n \geq 1$. For $\nu \in \boldsymbol{P}(S)$, define

$$\begin{aligned} H(\nu \mid \mu) &= -\int f\, lnf\, d\mu, \text{ if } \nu \ll \mu \text{ and } f = d\nu/d\mu, \\ &= \infty \text{ otherwise.} \end{aligned}$$

Clearly, $H(\mu \mid \mu) = 0$ and by Jensen's inequality, $H(\nu \mid \mu) \geq 0$.

Theorem 4.5.2 *(Sanov)* $\{\eta_n, n \geq 1\}$ *satisfies the large deviations principle with rate function* $H(. \mid \mu)$.

See [13, Ch. III] for a proof.

What's presented here are mere rudiments of large deviations. There is a vast edifice built upon these general ideas, with several layers of abstractions. The reader is encouraged to look up [13, 17, 46] for further results on large deviations.

4.6 Tests for Convergence

We shall study here some tests for convergence or nonconvergence of sums of independent random variables in the almost sure sense. (Recall that they converge with probability zero or one, by virtue of Kolmogorov's zero-one law.) The first result given below replaces this by apparently weaker convergence concepts (Theorem 4.6.1 below.). We shall need the following lemma:

Lemma 4.6.1 *Let* $\{X_n\}$ *be independent real-valued random variables and* $S_n = \sum_{i=1}^n X_i$, $n \geq 1$. *Let* $c > 0$. *Then*

$$P(max_{m<k\leq n} \mid S_k - S_m \mid > 2c) \leq \frac{P(\mid S_n - S_m \mid > c)}{(1 - max_{m<k\leq n} P(\mid S_n - S_k \mid \geq c))}.$$

Proof Let $\tau = \inf\{k > m \mid | S_k - S_m |> 2c\}(= \infty$ if the set in question is empty). Then

$$\begin{aligned}
&P(\max_{m<k\leq n} | S_k - S_m |> 2c) \\
&= \sum_{k=m+1}^{n} P(\tau = k, | S_k - S_m |> 2c) \\
&= \sum_{k=m+1}^{n} P(\tau = k, | S_k - S_m |> 2c, | S_n - S_k |\geq c) \\
&\quad + \sum_{k=m+1}^{n} P(\tau = k, | S_k - S_m |> 2c, | S_n - S_k |< c) \\
&\leq (\sum_{k=m+1}^{n} P(\tau = k, | S_k - S_m |> 2c))(\max_{m<k\leq n} P(| S_n - S_k |\geq c)) \\
&\quad + \sum_{k=m+1}^{n} P(\tau = k, | S_n - S_m |> c) \\
&\leq P(\max_{m<k\leq n} | S_k - S_m |> 2c)(\max_{m<k\leq n} P(| S_n - S_k |\geq c)) \\
&\quad + P(| S_n - S_m |> c).
\end{aligned}$$

The claim follows. □

Theorem 4.6.1 *(Lévy) The following are equivalent:*
(a) $\{S_n\}$ *converges a.s.*
(b) $\{S_n\}$ *converges in probability.*
(c) $\{S_n\}$ *converges in law.*

Proof Clearly (a) ⇒ (b) ⇒ (c). Suppose $S_n \to S_\infty$ in probability. Pick $\{n(k)\} \subset \{n\}$ such that for $k \geq 1$,

$$P(| S_i - S_{n(k)} |> 1/2^k) < 1/2^k, i \geq n(k).$$

Then

$$\sum_{k=1}^{\infty} P(| S_{n(k+1)} - S_{n(k)} |> 2^{-k}) < \infty$$

and thus, by Borel–Cantelli lemma, $P(| S_{n(k+1)} - S_{n(k)} |> 2^{-k}\ i.o.) = 0$ and $S_{n(k)}$ converges a.s. as $k \to \infty$. By the preceding lemma,

$$P(\max_{n(k)<i\leq n(k+1)} | S_i - S_{n(k)} |> 1/2^{k-1}) \leq 1/2^{k-1}$$

for $k \geq 1$. Using the Borel–Cantelli lemma as above,

$$P(\max_{n(k)<m\le n(k+1)} \mid S_m - S_{n(k)} \mid > 1/2^{k-1} \; i.o.) = 0.$$

Thus $S_m \to S_\infty$ a.s. and therefore (b) $\Rightarrow$ (a).

Next, suppose that $\mu_n =$ the law of $S_n, n \ge 1$, converge to μ in $\boldsymbol{P}(\boldsymbol{R})$ as $n \to \infty$. Let $\mu_{mn} =$ the law of $S_n - S_m$ for $n > m$. Since $\{\mu_n\}$ is tight, for any $\epsilon > 0$, there exists a compact $K_\epsilon \subset \boldsymbol{R}$ such that $\mu_n(K_\epsilon) > 1 - \epsilon$ for all n. Let $K_1 = \{x - y \mid x, y \in K_\epsilon\}$. Then K_1 is compact. Since $S_n - S_m \in K_1$ whenever $S_n, S_m \in K_\epsilon, m < n$,

$$\begin{aligned}\mu_{mn}(K_1) &\ge P(S_n \in K_\epsilon, S_m \in K_\epsilon)\\ &\ge 1 - P(S_n \in K_\epsilon^c) - P(S_m \in K_\epsilon^c)\\ &\ge 1 - 2\epsilon.\end{aligned}$$

Thus $\{\mu_{mn}, n > m, m = 1, 2, ...\}$ is tight. We need to show that $\{S_n\}$ is Cauchy in probability, that is, for any $\epsilon > 0$, there exists an $N_\epsilon \ge 1$ such that for all $n > m \ge N_\epsilon$,

$$\mu_{mn}((-\epsilon, \epsilon)) > 1 - \epsilon.$$

Suppose not. Then there exists an $\epsilon > 0$ such that for all $N \ge 1$, there exist $n(N) > m(N) > N$ for which

$$\mu_{m(N)n(N)}((-\epsilon, \epsilon)) \le 1 - \epsilon.$$

Let ν be a limit point of $\{\mu_{m(N)n(N)}\}$ along some subsequence denoted by, say, $\{\mu_{m(N_i)n(N_i)}\}$. Then

$$\nu((-\epsilon, \epsilon)) \le \liminf_{i\to\infty} \mu_{m(N_i)n(N_i)}((-\epsilon, \epsilon)) \le 1 - \epsilon.$$

Letting $\psi_n(.)$ denote the characteristic function of μ_n for $n \ge 1$ and $\psi_{mn}(.)$ that of $\mu_{mn}, n > m, m \ge 1$,

$$\begin{aligned}\psi_{n(N_i)}(t) &= E[\exp(itS_{n(N_i)})]\\ &= E[exp(itS_{m(N_i)})]E[exp(it(S_{n(N_i)} - S_{m(N_i)}))]\\ &= \psi_{m(N_i)}(t)\psi_{m(N_i)n(N_i)}(t)\end{aligned}$$

for $t \in \boldsymbol{R}$. Letting $i \to \infty, \varphi_\mu(t) = \varphi_\mu(t)\varphi_\nu(t), t \in \boldsymbol{R}$, where φ_μ, φ_ν are characteristic functions of μ, ν respectively. Since $\varphi_\mu(0) = 1$ and $\varphi_\mu(.)$ is continuous, $\varphi_\mu(t) \ne 0$ on $\{-r \le t \le r\}$ for some $r > 0$. Thus $\varphi_\nu(t) = 1$ for $-r \le t \le r$. Using the inequality $\mid \varphi_\nu(t) - \varphi_\nu(s) \mid \le \sqrt{2 \mid 1 - \varphi_\nu(t-s) \mid}$ (cf. Exercise 2.17), we get $\varphi_\nu(.) = 1$. Thus $\nu =$ the Dirac measure at 0, implying $1 = \nu((-\epsilon, \epsilon)) \le 1 - \epsilon$, a contradiction. Thus (c) $\Rightarrow$ (b). □

The following characterization of the a.s. convergence of $\{S_n\}$ often serves as a useful test thereof. Let $\sigma^2(...)$ stand for "the variance of ...".

Theorem 4.6.2 *(Kolmogorov's three series theorem) Let $\{X_n\}, \{S_n\}$ be as above, $c > 0$ and $Y_n = X_n I\{| X_n |\leq c\}, n \geq 1$. Then S_n converges a.s. as $n \to \infty$ if and only if the following three series converge:*
(i) $\sum_{n=1}^{\infty} P(X_n \neq Y_n)$,
(ii) $\sum_{n=1}^{\infty} E[Y_n]$,
(iii) $\sum_{n=1}^{\infty} \sigma^2(Y_n)$.

Proof Suppose (i) – (iii) converge. Since (iii) converges, Theorem 3.3.4 applied to the martingale $(S'_n - E[S'_n], \sigma(X_i, i \leq n)), n \geq 1$, where $S'_n = Y_1 + ... + Y_n, n \geq 1$, implies that $\{S'_n - E[S'_n]\}$ converges a.s. (Exercise 4.15). Since (ii) converges, $\{E[S'_n]\}$ converges and therefore so does $\{S'_n\}$, a.s. Since (i) converges, the Borel–Cantelli lemma implies that $X_n = Y_n$ from some n on, a.s. Thus $\{S_n\}$ must converge a.s.

Conversely, suppose that $\{S_n\}$ converges a.s. Then $X_n \to 0$ a.s. and also $| X_n |\leq c$ for $n \geq n_0(\omega)$ for some integer valued random variable n_0 satisfying $n_0 < \infty$ a.s. Thus $X_n = Y_n$ from some n on, a.s. By the Borel–Cantelli lemma, (i) must converge. Also, the foregoing implies that $\{S'_n\}$ converges a.s. Let $s_n =$ the nonnegative square-root of $\sum_{m=1}^{n} \sigma^2(Y_m), n \geq 1$. Suppose $s_n \to \infty$. By the central limit theorem, $(S'_n - E[S'_n])/s_n \to N(0,1)$ in law. Since $S'_n/s_n \to 0$ a.s., we must have $E[S'_n]/s_n \to N(0,1)$ in law, a contradiction (Exercise 4.16). Thus $\{s_n\}$ and therefore (iii) converges. Invoking Theorem 3.3.4 again, we conclude that $\{S'_n - E[S'_n]\}$ converges a.s. Thus (ii) must converge, completing the proof. □

The final result we shall consider here reduces the problem concerning random variables $\{X_n\}$ to one concerning their "symmetrized" counterparts which are sometimes easier to handle. First construct (by augmenting the underlying probability space if necessary) real random variables $\{Y_n\}$ such that $(X_n, Y_n, n \geq 1)$ is an independent family and X_n, Y_n agree in law for each n. Let $Z_n = X_n - Y_n, n \geq 1$. Then each Z_n is "symmetric" in the sense that $P(Z_n \leq -x) = P(Z_n \geq x)$ for all x. This is what we mean by a "symmetrized counterpart of X_n". Let $T_n = \sum_{i=1}^{n} Z_i, V_n = \sum_{i=1}^{n} Y_i, n \geq 1$. Then $T_n = S_n - V_n$ for $n \geq 1$.

Theorem 4.6.3 *$\{T_n\}$ converges a.s. if and only if there exist $\{b_m\} \subset \boldsymbol{R}$ such that $\{S_n - \sum_{m=1}^{n} b_m\}$ converges a.s.*

Proof If the latter converges a.s., so will $V_n - \sum_{m=1}^{n} b_m, n \geq 1$. Since $T_n = (S_n - \sum_{m=1}^{n} b_m) - (V_n - \sum_{m=1}^{n} b_m), n \geq 1, \{T_n\}$ must converge a.s. Conversely, suppose $\{T_n\}$ converges a.s. Fix $n \geq 1$ and $\epsilon > 0$. Pick $a_n \in \boldsymbol{R}$ such that $P(X_n \geq a_n) \geq 1/2, P(X_n \leq a_n) \geq 1/2$. Then $P(Y_n \geq a_n) \geq 1/2, P(Y_n \leq a_n) \geq 1/2$ as well. Thus

$$P(Z_n \geq \epsilon) \quad = \quad P((X_n - a_n) - (Y_n - a_n) \geq \epsilon)$$

$$\begin{aligned} &\geq P(X_n - a_n \geq \epsilon, Y_n - a_n \leq 0) \\ &= P(X_n - a_n \geq \epsilon)P(Y_n - a_n \leq 0) \\ &\geq \frac{1}{2}P(X_n - a_n \geq \epsilon). \end{aligned}$$

Similarly one proves

$$P(Z_n \leq -\epsilon) \geq \frac{1}{2}P(X_n - a_n \leq -\epsilon).$$

Thus

$$P(| Z_n |\geq \epsilon) \geq \frac{1}{2}P(| X_n - a_n | \geq \epsilon),$$

implying, for $c > 0$,

$$\begin{aligned} E[| X_n - a_n |^2 I\{| X_n - a_n |\leq c\}] &= 2\int_0^c xP(| X_n - a_n |\geq x)dx \\ &\leq 4\int_0^c xP(| Z_n |\geq x)dx \\ &= 2E[Z_n^2 I\{| Z_n |\leq c\}]. \qquad (4.8) \end{aligned}$$

Since Z_n is symmetric, so is $Z_n I\{| Z_n |\leq c\}$, implying in particular that it is zero mean. Thus the right hand side of (4.8) equals $2\sigma^2(Z_n I\{| Z_n |\leq c\})$. Since $\{T_n\}$ converges a.s., Theorem 4.6.2 implies that

$$\sum_{n=1}^{\infty} \sigma^2(Z_n I\{| Z_n |\leq c\}) < \infty,$$

and thus

$$\sum_{n=1}^{\infty} E[| X_n - a_n |^2 I\{| X_n - a_n |\leq c\}] < \infty.$$

Therefore

$$\sum_{n=1}^{\infty} \sigma^2((X_n - a_n)I\{| X_n - a_n |\leq c\}) < \infty.$$

Using Theorem 3.3.4 as in the proof of the preceding theorem, it follows that

$$\sum_{m=1}^{n} ((X_m - a_m)I\{| X_m - a_m |\leq c\} - E[(X_m - a_m)I\{| X_m - a_m |\leq c\}])$$

converges a.s. as $n \to \infty$. Theorem 4.6.2 also implies that

$$\sum_{n=1}^{\infty} P(| Z_n |\geq c) < \infty.$$

Thus

$$\sum_{n=1}^{\infty} P(\mid X_n - a_n \mid \geq c) < \infty$$

and hence by the Borel–Cantelli lemma, $P(\mid X_n - a_n \mid \geq c \; i.o.) = 0$. It follows that $\{S_n - \sum_{m=1}^{n} b_m\}$ converges a.s. for $b_m = a_m + E[(X_m - a_m)I\{\mid X_m - a_m \mid \leq c\}], m \geq 1$. This completes the proof. □

For further results along these lines, see [31, Ch. V]. Some other references relevant to this chapter (in addition to those mentioned in the chapter) are [18, 23, 28].

4.7 Additional Exercises

(4.17) Let $\{X_i\}$ be i.i.d. real valued random variables with $E[X_i] = \infty, i \geq 1$. Show that $S_n/n \to \infty$ a.s.

(4.18) Let $\{X_i\}$ be pairwise independent, uniformly integrable real random variables with zero mean. Show that $S_n/n \to 0$ in probability.

(4.19) Show that for independent real random variables $\{X_n\}, X_n \to 0$ in probability need not imply $S_n/n \to 0$ in probability. (Hint: Let X_n take values 2^n and 0 with probability n^{-1} and $1 - n^{-1}$ resp.)

(4.20) Let $\{X_i\}$ be i.i.d. integrable real random variables with $E[X_1] \neq 0$. Show that $(\max_{1 \leq j \leq n} \mid X_j \mid)/ \mid S_n \mid \to 0$ a.s.

(4.21) If $\{X_n\}$ are i.i.d. zero mean real random variables, show that Lindberg's condition holds if for some $\delta > 0, (\sum_{i=1}^{n} E[\mid X_i \mid^{2+\delta}])/s_n^{2+\delta} \to 0$ as $n \to \infty$. (This sufficient condition is due to Liapunov.)

(4.22) Let $\{X_i\}$ be i.i.d. zero mean real random variables with $E[X_1^2] < \infty$. Show that

$$\begin{aligned} \liminf_{n \to \infty} \mid S_n \mid / \sqrt{n} &= 0 \text{ a.s.,} \\ \limsup_{n \to \infty} \mid S_n \mid / \sqrt{n} &= \infty \text{ a.s..} \end{aligned}$$

(4.23) For $I(.), J(.)$ as in Section 4.5, suppose that $I(x) < \infty, J(\theta) < \infty \; \forall x, \theta \in \boldsymbol{R}$. Show that

$$J(\theta) = \sup_x (\theta x - I(x)).$$

5

Markov Chains

5.1 Construction and the Strong Markov Property

A stochastic process or a random process is a family of random variables $X_t, t \in I$, taking values in some measurable space (E, ξ), with $I =$ an interval of integers or reals. I has the interpretation of a (discrete or continuous) time interval, t thereby being a time index. The simplest random process in discrete time is a sequence of independent random variables. This can be considered as a process without memory: the value of the process at a given time instant is independent of the values it took in the past. The next level of complication one can conceive of is that of a one step memory: the value taken by the process at time $n+1$ depends on its past values only through its dependence, if any, on the value at time n, not otherwise. To be precise, it is conditionally independent of its values up to time n conditioned on its value at time n. This is called the Markov property and the process a Markov chain.

At another level, it may be considered the probabilistic analog of a dynamical system. A (deterministic) dynamical system on a suitable "state space" entails a family of trajectories evolving in this space, satisfying certain properties, the principal one being that its evolution from some time t on is entirely dictated by its position at time t and not by how it arrived there. Compare this with the fact that the conditional law of the future trajectory of a Markov chain given the past trajectory is the same as that given its present value. A formal definition follows.

Let S be a countable set, called the "state space". Without loss of generality, we may label it {1,2,...}. Let $\lambda \in \boldsymbol{P}(S)$, and $P = [[p(i,j)]]_{i,j \in S}$ a matrix (infinite if S is so) satisfying: $p(i,j) \in [0,1], \sum_k p(i,k) = 1$ for all $i, j \in S$. Such a matrix is called a stochastic matrix. In particular, $p(i,.)$ can be identified with an element of $\boldsymbol{P}(S)$ for each i. The number $p(i,j)$ is called the transition probability of a transition from i to j. P is called the transition matrix. The reason for this suggestive terminology will be clear soon.

A sequence $\{X_n, n = 0,1,2,...\}$ of S-valued random variables is called a Markov chain with state space S, initial distribution λ and transition matrix P if
(a) $P(X_0 \in A) = \lambda(A), \forall A \subset S$,
(b) $P(X_{n+1} = j / X_n = i) = p(i,j), \forall i, j \in S, n \geq 0$,
(c) $P(X_{n+1} \in A/\mathcal{F}_n) = P(X_{n+1} \in A/X_n), \forall A \subset S, n \geq 0$,
where $\mathcal{F}_n = \sigma(X_0, X_1, ..., X_n), n \geq 0$.

Strictly speaking, $\{X_n\}$ above defines only a time-homogeneous Markov chain or a Markov chain with stationary transitions. We shall comment on this qualification later. Property (c) above is called the Markov property and easily generalizes (Exercise 5.1) to:
(c_1) $P((X_n, X_{n+1}, X_{n+2}, ...,) \in B/\mathcal{F}_n) = P((X_n, X_{n+1}, X_{n+2}, ...) \in B/X_n)$ for $n \geq 0, B \in$ the product σ-field of $S^\infty = S \times S \times \cdots$.

As mentioned at the beginning, we identify the subscript n above with discrete time steps. Then (c_1), in view of Theorem 3.1.2, is seen to be equivalent to the intuitively appealing statement: the *future* $(X_{n+1}, X_{n+2}, ...)$ and the *past* $(X_0, ..., X_{n-1})$ are conditionally independent given the *present*, i.e., X_n, for each n. Since this statement is symmetric in time, (c_1) is also equivalent to
(c_2) $P((X_0, X_1, ..., X_n) \in B/X_n, X_{n+1}, ...) = P((X_0, X_1, ..., X_n) \in B/X_n)$, for $n \geq 0, B \subset S^{n+1}$.

Given S, λ, P as above, one can construct a Markov chain $\{X_n\}$ satisfying (a) - (c) above on the "canonical" space $\Omega = S^\infty$ endowed with the product σ-field $\mathcal{F}$. One proceeds as follows: Let $A \in \mathcal{F}$ be of the form $A = \Pi_{i=0}^\infty A_i$ where $A_i \subset S$ and $A_i \neq S$ for at most finitely many i. Define a probability measure P on $(\Omega, \mathcal{F})$ by

$$P(A) = \sum_{j_0 \in A_0} \sum_{j_1 \in A_1} \cdots \sum_{j_n \in A_n} \lambda(j_0) p(j_0, j_1) p(j_1, j_2) ... p(j_{n-1}, j_n)$$

for any $n \geq 1$ for which $i > n$ implies $A_i = S, \lambda(j)$ being simply $\lambda(\{j\})$. It is easily checked that this definition is independent of the specific choice of n and leads to a consistent family of probability measures on $S^n, n \geq 0$ (Exercise 5.2). By Kolmogorov extension theorem, P extends uniquely to a probability measure on $(\Omega, \mathcal{F})$. For $n \geq 0$ and $\omega = (\omega_0, \omega_1, \omega_2, ...) \in \Omega$, define $X_n(\omega) = \omega_n$. Then $\{X_n\}$ is the desired Markov chain. From now on, we assume this canonical set-up.

It is clear that the law of $\{X_n\}$ is uniquely specified by λ and P. Let $P_j \in \boldsymbol{P}(\Omega)$ denote the law of $(X_0, X_1, X_2, ...)$ when λ is the Dirac measure at $j \in S$. Then (c_1) can be rewritten as

(c_3) $P((X_n, X_{n+1}, ...) \in B/\mathcal{F}_n) = P_{X_n}((X_0, X_1, ...) \in B), n \geq 0, B \in \mathcal{F}.$

$\{X_n\}$ is said to have the strong Markov property (SMP for short) if for any $\{\mathcal{F}_n\}$-stopping time τ and $B \in \mathcal{F}$,

$$I\{\tau < \infty\}P((X_\tau, X_{\tau+1}, ...) \in B/\mathcal{F}_\tau) = I\{\tau < \infty\}P_{X_\tau}((X_0, X_1, ...) \in B)\text{a.s.}$$

Theorem 5.1.1 *$\{X_n\}$ has the SMP.*

Proof It suffices to show that for all $A \in \mathcal{F}_\tau$,

$$\begin{aligned}\int_A & I\{\tau < \infty\}P((X_\tau, X_{\tau+1}, ...) \in B/\mathcal{F}_\tau)dP \\ &= \int_A I\{\tau < \infty\}P_{X_\tau}((X_0, X_1, ...) \in B)dP. \end{aligned} \tag{5.1}$$

Let $\varphi(j) = P_j((X_0, X_1, ...) \in B), j \in S$, for a prescribed $B \in \mathcal{F}$. Then the left hand side of (5.1) equals

$$\begin{aligned} P(\{(X_\tau, & X_{\tau+1}, ...) \in B\} \cap A \cap \{\tau < \infty\}) \\ &= \sum_n P(\{(X_n, X_{n+1}, ...) \in B\} \cap A \cap \{\tau = n\}) \\ &= \sum_n E[P((X_n, X_{n+1}, ...) \in B/\mathcal{F}_n)I_{A\cap\{\tau=n\}}] \\ &= \sum_n \int_{A\cap\{\tau=n\}} \varphi(X_n)dP \qquad \text{...by } (c_1) \\ &= \int_{A\cap\{\tau<\infty\}} \varphi(X_\tau)dP \end{aligned}$$

which equals the right hand side of (5.1). □

The statement of the SMP can be further simplified as follows: Let $\theta_n : S^\infty \to S^\infty$ denote the measurable map $(\omega_0, \omega_1, \omega_2, ...) \to (\omega_n, \omega_{n+1}, ...), n \geq 0$. Then $X_m \circ \theta_n = X_{m+n}$ for all m, n, in the canonical set-up. For an $\{\mathcal{F}_n\}$-stopping time τ, let $\theta_\tau(\omega) = \theta_{\tau(\omega)}(\omega)$ for $\omega \in \{\tau < \infty\}$. Then SMP is equivalent to either of the following two statements (Exercise 5.3): For any $\{\mathcal{F}_n\}$ stopping time τ, either

(i) $P(A \bigcap \theta_\tau^{-1}(B)) = P(A)P_j(B)$ for all $A \in \mathcal{F}_\tau$ satisfying $A \in \{\tau < \infty\} \bigcap \{X_\tau = j\}$, all $B \in \mathcal{F}$ and $j \in S$, or,

(ii) $E[Y \circ \theta_\tau / \mathcal{F}_\tau] = E_{X_\tau}[Y]$ a.s. on $\{\tau < \infty\}$, for $Y \in L_1(\Omega, \mathcal{F}, P), E_j[\,\cdot\,]$ being the expectation under P_j.

5.2 Classification of States

In this section, we classify the elements of S according to certain recurrence properties to be defined.

Let $\tau_j = min\{n > 0 \mid X_n = j\}$ for $j \in S$ denote the "first hitting time" of j ("first return time" if $X_0 = j$). Define $\rho_{ij} = P_i(\tau_j < \infty), N_j = \sum_{n=1}^{\infty} I\{X_n = j\}$ and $G(i,j) = E_i[N_j]$. (N_j and $G(i,j)$ can be $+\infty$.) Letting P^n denote the n times matrix product of P with itself and $p^n(i,j)$ its $(i,j)-$th element, we have

$$G(i,j) = \sum_{n=1}^{\infty} p^n(i,j).$$

Lemma 5.2.1 $P_i(N_j = m) = \rho_{ij}\rho_{jj}^{m-1}(1-\rho_{jj}),\ m \geq 1.$

Proof Let $\sigma_1 = \tau_j, \sigma_n = min\{m > \sigma_{n-1} \mid X_m = j\}$ for $n = 2, 3, \ldots$. Then

$$\begin{aligned} P_i(N_j = m) &= P_i(\sigma_l < \infty, 1 \leq l \leq m, \sigma_{m+1} = \infty) \\ &= P_i(\tau_j < \infty)\underbrace{P_j(\tau_j < +\infty)\ldots P_j(\tau_j < \infty)}_{(m-1)\ \text{times}} P_j(\tau_j = \infty) \\ &= \rho_{ij}\rho_{jj}^{m-1}(1-\rho_{jj}) \end{aligned}$$

where the second step follows by repeated application of the SMP. □

Theorem 5.2.1 *If* $\rho_{jj} < 1, P_i(N_j = \infty) = 0$ *and* $G(i,j) < \infty$. *If* $\rho_{jj} = 1, P_j(N_j = \infty) = 1, G(j,j) = \infty, P_i(N_j = \infty) = \rho_{ij} = 1 - P_i(N_j = 0)$ *and* $G(i,j) = \infty$ *or* 0 *according to whether* $\rho_{ij} > 0$ *or* $= 0$.

Proof This is immediate from the preceding lemma (Exercise 5.4). □

Let $S_T = \{j \in S \mid \rho_{jj} < 1\}$ and $S_R = \{j \in S \mid \rho_{jj} = 1\}$. The elements of S_T and S_R are called transient and recurrent states respectively. Write $i \to j$ whenever $\rho_{ij} > 0$.

Theorem 5.2.2 *If* $i \in S_R$ *and* $i \to j$, *then* $j \in S_R$ *and* $\rho_{ij} = \rho_{ji} = 1$.

Proof If $\rho_{ji} < 1, P_j(X_m \neq i, m \geq 0) > 0$. Together with the fact that $i \to j$, this implies that $\rho_{ii} < 1$, that is, $i \notin S_R$, a contradiction. This implies that $\rho_{ji} = 1$. Let n_1, n_2 be integers satisfying $p^{n_1}(j,i) > 0, p^{n_2}(i,j) > 0$. Then $\alpha = p^{n_1}(j,i)p^{n_2}(i,j) > 0$. Note that

$$p^{n_1+n+n_2}(j,j) \geq \alpha p^n(i,i).$$

Summing over n and using the fact that $G(i,i) = \infty$, we have $G(j,j) = \infty$ and thus $j \in S_R$. By the symmetry of the first step, $\rho_{ij} = 1$. □

It follows that on $S_R, i \to j$ is an equivalence relation. $E \subset S$ is said to be closed if $p(i,E) = 1$ for any $i \in E$. It is said to be irreducible if it is closed, and furthermore $i \to j$ for $i,j \in E$. The Markov chain itself is said to be irreducible if S is. By the above theorem, it is then either transient (i.e. $S = S_T$) or recurrent (i.e., $S = S_R$).

Define $N_n(j) = \sum_{m=1}^n I\{X_m = j\}, G_n(i,j) = E_i[N_n(j)], T_j = E_j[\tau_j]$ (possibly $+\infty$) and $\tau_j^n = R_j^n - R_j^{n-1}$ for $n \geq 1$ where $\{R_j^n\}$ are defined recursively as

$$R_j^0 = \tau_j, R_j^n = min\{n > R_j^{n-1} \mid X_n = j\}, n \geq 1.$$

Theorem 5.2.3 *(a) $j \in S_T$ implies: $N_n(j)/n \to 0, P_i$-a.s., $G_n(i,j)/n \to 0$ for all $i \in S$.*
(b) $j \in S_R$ implies: $N_n(j)/n \to T_j^{-1}I\{\tau_j < \infty\}, P_i$-a.s. and $G_n(i,j)/n \to \rho_{ij}T_j^{-1}$ where by convention, $a/\infty = 0$ for any $a \in \boldsymbol{R}$.

Proof (a) Since $N_n(j) \leq N_j < \infty, P_i$- a.s. and $G_n(i,j) \leq G(i,j) < \infty$, the claim follows immediately.
(b) Let $i = j$. By SMP, $\{\tau_j^n, n \geq 1\}$ are i.i.d. By the strong law of large numbers,

$$R_j^n/n = n^{-1}\sum_{m=1}^{n} \tau_j^m \to T_j, P_j - a.s.$$

Since $R_j^{N_n(j)} \leq n < R_j^{N_n(j)+1}$, the first claim follows for $i = j$. For $i \neq j$, apply the same argument to the Markov chain $\{X_{\tau_j+m}, m \geq 0\}$ defined on the probability space $(\Omega', \mathcal{F}', P')$ where $\Omega' = \{\tau_j < \infty\}, \mathcal{F}' =$ the corresponding trace σ-field and $P' = P$ restricted to $\{\tau_j < \infty\}$ and renormalized to make it a probability measure. (Of course, we assume that $P(\tau_j < \infty) > 0$, the other case being trivial.) The first half of (b) follows. The claim concerning $G_n(i,j)$ now follows by the dominated convergence theorem.

□

Let $S_P = \{j \in S_R \mid T_j < \infty\}$ and $S_N = \{j \in S_R \mid T_j = \infty\}$. The elements of S_P and S_N are known respectively as the positive recurrent and the null recurrent states.

Theorem 5.2.4 *If $i \in S_P$ and $i \to j$, then $j \in S_P$.*

Proof Since $i \to j, j \to i$. Let n_1, n_2 be integers such that $p^{n_1}(j,i) > 0, p^{n_2}(i,j) > 0$. Then for $\alpha = p^{n_1}(j,i)p^{n_2}(i,j) > 0$, we have

$$p^{n_1+n+n_2}(j,j) \geq \alpha p^n(i,i).$$

Hence

$$\begin{aligned}\frac{1}{n}\sum_{m=1}^{n} p^{n_1+n+n_2}(j,j) &= \frac{G_{n_1+n+n_2}(j,j)}{n} - \frac{G_{n_1+n_2}(j,j)}{n} \\ &\geq \alpha\frac{G_n(i,i)}{n} - \frac{G_{n_1+n_2}(j,j)}{n}.\end{aligned}$$

Letting $n \to \infty$, the left-hand side and the right-hand side converge respectively to T_j^{-1} and αT_i^{-1}, implying $T_j^{-1} > 0$, that is, $T_j < \infty$. □

It follows that $i \to j$ is an equivalence relation on S_P and S_N. Thus S can be written as the disjoint union $S = S_T \cup S_P \cup S_N$, where each $i \in S_P$ or S_N lies in an equivalence class with respect to the above equivalence relation. Each such class, called a communicating class, is obviously closed. Also, it is clear that $p^n(i,j) = 0, n \geq 0$, when i, j belong to two distinct equivalence classes of the above type and thus

$$n^{-1}\sum_j G_n(i,j) = 1, n \geq 1, \tag{5.2}$$

where j is summed over such an equivalence class containing i. If this equivalence class is finite, Theorems 5.2.3(b) and (5.2) lead to

$$\sum_j T_j^{-1} = 1$$

where the summation is as above. Thus at least one of the T_j's is finite, implying that the corresponding j is in S_P. Thus the equivalence class itself must be in S_P if it is finite. If it is a singleton, say $\{i\}, i$ is called an absorbing state and must satisfy

$$p(i,j) = \delta_{ij} \text{ (the Kronecker delta)}, \quad j \in S.$$

Combining Theorems 5.2.1 - 5.2.4 above, we can now describe the qualitative behaviour of a general Markov chain. A Markov chain starting at $i \in S_R$ remains in the communicating class containing i, and visits each state in this class infinitely often with probability one. The mean return time to state i, $E_i[\tau_i]$, is finite if $i \in S_P$ and infinite if $i \in S_N$. If it starts at some $i \in S_T$, then there are two possibilities. One is that it can end up in a finite (random) time in one of the equivalence classes of S_R described above and remain in it thereafter. The second possibility is that it never hits S_R, but drifts to infinity. That is, for each finite subset of S, there is a finite random time after which the chain does not visit this set. If S_T is finite, only the first possibility exists.

The Markov chain can be represented by a weighted digraph (directed graph) as follows: Let each $i \in S$ correspond to a node of the graph, and draw an edge directed from i to j if $p(i,j) > 0$. Assign a weight $p(i,j)$ to this edge. Thus at each node, weights of the outgoing edges add up to one. A node corresponds to an absorbing state if and only if the only outgoing edge from the node returns to itself. The relation $i \sim j$ on the nodes defined by "a directed path exists from i to j and from j to i" is an equivalence relation, and corresponds to $i \to j, j \to i$ above. A closed equivalence class under this definition will be called a communicating class. This will correspond to the concept of a communicating class defined earlier if it lies in S_R. Note, however, that with this new definition, a communicating class can also be transient. Graphically, a communicating class is characterized as follows: A is a communicating class if no edges leave A and for any $i, j \in A$, directed paths exist from i to j and from j to i. Note also that recurrence and positive recurrence of a communicating class of infinite cardinality do not depend only on its directed graph, but also depend on the weights we assign to its edges. Finally, the chain is irreducible if S is a single communicating class. An irreducible Markov chain is said to be positive recurrent, null recurrent or transient according to whether $S = S_P$, S_N or S_T.

5.3 Stationary Distributions

A measure π on S is said to be stationary or invariant for the Markov chain $\{X_n\}$ if

$$\sum_{j \in S} p(j,k)\pi(j) = \pi(k), \; k \in S,$$

where $\pi(i) = \pi(\{i\})$ for $i \in S$. Writing π as a row vector $[\pi(1), \pi(2), ...]$, the above becomes $\pi P = \pi$ in matrix notation. Thus $\pi P^n = \pi$ for $n \geq 1$ and hence $\pi G_n/n = \pi, n \geq 1$, where $G_n = [[G_n(i,j)]]$. If π is a probability measure in addition, we call it a stationary distribution or an invariant probability measure. It is clear that if π is an invariant probability measure and the law of X_0 is π, then the law of X_n will be π for all n. In fact, the laws of $(X_0, X_1, X_2, ...)$ and $(X_n, X_{n+1}, X_{n+2}, ...)$ agree for all n, making $\{X_n\}$ a "stationary process".

Theorem 5.3.1 *If π is an invariant probability measure for $\{X_n\}$, then $\pi(i) = 0$ for all $i \notin S_P$.*

Proof For $i \notin S_P, G_n(j,i)/n \to 0$. By the dominated convergence theorem, the i-th component of $\pi G_n/n$ tends to zero as $n \to \infty$. But $\pi G_n/n = \pi$. The claim follows. □

Theorem 5.3.2 *Let $\{X_n\}$ be irreducible and positive recurrent (that is, $S = S_P$). Then there exists a unique stationary distribution $\pi \in \boldsymbol{P}(S)$ given by*

$$\pi(i) = T_i^{-1},\ i \in S.$$

Furthermore, under any initial law and for any $f \in C_b(S)$,

$$\frac{1}{n}\sum_{m=0}^{n-1} f(X_m) \to \sum_i \pi(i)f(i),\ a.s.$$

Proof If π is a stationary distribution, then

$$G_n(i,j)/n \to T_j^{-1}$$

and

$$\pi = \pi G_n/n$$

together imply (by the dominated convergence theorem) that $\pi(i) = T_i^{-1}$, $i \in S$. Thus it suffices to show that $\pi(i) = T_i^{-1}, i \in S$, defines a stationary distribution. Let S be finite. Since

$$\sum_{j\in S} G_n(i,j)/n = 1,$$

one has

$$1 = \lim_{n\to\infty}\sum_{j\in S} G_n(i,j)/n = \sum_{j\in S}\lim_{n\to\infty} G_n(i,j)/n = \sum_{j\in S} T_j^{-1}.$$

Since $G_nP = G_{n+1} - P$,

$$\begin{aligned} T_j^{-1} &= \lim_{n\to\infty} G_n(i,j)/n \\ &= \lim_{n\to\infty}\sum_{k\in S} G_n(i,k)p(k,j)/n \\ &= \sum_{k\in S}\lim_{n\to\infty} G_n(i,k)p(k,j)/n \\ &= \sum_{k\in S} T_k^{-1}p(k,j) \end{aligned} \tag{5.3}$$

for $j \in S$, proving the desired result. If S is infinite, take $S_1 \subset S$ finite. Let $n \to \infty$ in the inequality

$$\sum_{j\in S_1} G_n(i,j)/n \le 1$$

to conclude that

$$\sum_{j\in S_1} T_j^{-1} \le 1.$$

Similarly, we argue as for (5.3) to conclude

$$\sum_{j\in S_1} T_j^{-1} p(j,k) \le T_k^{-1}, k \in S.$$

Thus

$$\begin{aligned} \sum_{j\in S} T_j^{-1} &= c^{-1} \le 1, \\ \sum_{j\in S} T_j^{-1} p(j,k) &\le T_k^{-1}, k \in S. \end{aligned} \tag{5.4}$$

Summing (5.4) over k, we get,

$$\sum_{j\in S} T_j^{-1} \le \sum_{k\in S} T_k^{-1}.$$

Thus we must have equality in (5.4), implying that $\pi(i) = cT_i^{-1}, i \in S$, defines a stationary distribution. From the first part of the proof, it follows that $c = 1$. For the second claim, consider the empirical measures $\mu_n \in \boldsymbol{P}(S)$ defined by

$$\int f d\mu_n = \frac{1}{n} \sum_{m=0}^{n-1} f(X_m), \; f \in C_b(S), \; n \ge 1.$$

Then by Theorem 5.2.3(b) and the foregoing $\mu_n(\{i\}) \to \pi(i)$ a.s., $i \in S$. By Theorem 2.3.3, $\mu_n \to \pi$ a.s. in total variation. Thus $\int f d\mu_n \to \int f d\pi$ a.s. for $f \in C_b(S)$. □

Let $\Gamma \subset \boldsymbol{P}(S)$ denote the set of stationary distributions for $\{X_n\}$. If $S_P = \phi, \Gamma = \phi$ by Theorem 5.3.1. If $S_P \ne \phi$, support$(\pi) \subset S_P$ for all $\pi \in \Gamma$. Let $C_1, C_2, ...$ be the (possibly infinitely many) communicating classes in S_P. They are mutually disjoint. Let $P^{(i)} = [[p(j,k)]]$ for $j,k \in C_i$. Then $P^{(i)}$ will be the transition matrix of an irreducible positive recurrent Markov chain on C_i. By the above theorem, there is a unique $\pi_i \in \boldsymbol{P}(C_i)$ such that $\pi_i P^{(i)} = \pi_i, i = 1, 2, \ldots$. We may consider π_i as an element of $\boldsymbol{P}(S)$ by setting $\pi_i(j) = 0$ for $j \notin C_i$. On the other hand for any $\pi \in \Gamma$ such that

$$a_i = \sum_{j\in C_i} \pi(j) > 0,$$

the measure $\tilde{\pi}_i \in \boldsymbol{P}(C_i)$ defined by

$$\tilde{\pi}_i(j) = \pi(j)/a_i, \; j \in C_i,$$

satisfies $\tilde{\pi}_i P^{(i)} = \tilde{\pi}_i$ and hence coincides with π_i. Thus any $\pi \in \Gamma$ may be written as $\pi = \sum a_i \pi_i$ where

$$a_i = \sum_{j \in C_i} \pi(j),\ i = 1, 2,$$

satisfy

$$a_i \geq 0,\ \sum_i a_i = 1.$$

Conversely, any convex combination of $\{\pi_i\}$ is in Γ, as is easily verified. Thus we have

Corollary 5.3.1 $\Gamma = \{\pi \in \boldsymbol{P}(S) \mid \pi = \sum_i a_i \pi_i, a_i \geq 0, \sum_i a_i = 1\}$, *where* $\{\pi_i\}$ *are as above, viewed as elements of* $\boldsymbol{P}(S)$ *by assigning zero mass to the complement of the corresponding* C_i.

For $i \in S$ satisfying $\rho_{ii} > 0$ (i.e., $p^n(i,i) > 0$ for some $n \geq 1$), define the period of i, denoted d_i, to be the greatest common divisor of the set $\{n > 1 \mid p^n(i,i) > 0\}$.

Lemma 5.3.1 *If* $i \to j$ *and* $j \to i, d_i = d_j$.

Proof Take $n_1 \geq 1, n_2 \geq 1$ such that

$$p^{n_1}(i,j) > 0,\ p^{n_2}(j,i) > 0.$$

Thus

$$p^{n_1+n_2}(i,i) > 0.$$

Therefore d_i divides $n_1 + n_2$. For any n satisfying

$$p^n(j,j) > 0,$$

we have

$$p^{n_1+n+n_2}(i,i) > 0,$$

implying that d_i divides $n_1 + n + n_2$. Thus d_i must divide n, implying $d_i \leq d_j$. A symmetric argument gives $d_j \leq d_i$. □

An irreducible Markov chain is said to be periodic with period d if $d_i = d > 1$ for all i, and aperiodic if $d_i = 1$ for all i. (Recall from the above lemma that d_i will be independent of i.)

Theorem 5.3.3 *Let* $\{X_n\}$ *be irreducible and positive recurrent with stationary distribution* π.

(a) *If* $\{X_n\}$ *is aperiodic,* $\lim_{n\to\infty} p^n(i,j) = \pi(j),\ i,j \in S$.

(b) If $\{X_n\}$ is periodic with period d, then for each pair (i,j) in $S \times S$, there exists an integer r with $0 \leq r < d$ such that $p^n(i,j) = 0$ unless $n = md + r$ for some $m \geq 0$ and

$$\lim_{m\to\infty} p^{md+r}(i,j) = d\pi(j).$$

The proof will be based on the following lemma:

Lemma 5.3.2 *Let $I \subset \mathbf{N}_0 = \{0,1,2,...\}$ satisfy: (i) $I + I \subset I$, (ii) the greatest common divisor of I is 1. Then there exists an $n_0 \in \mathbf{N}_0$ such that $n \in I$ for $n \geq n_0$.*

Proof We claim that there exists an $n_1 \in \mathbf{N}_0$ such that both n_1 and $n_1 + 1$ are in I. If not, there exists $k \geq 2$ in $\mathbf{N}_0$ and $n_1 \in I$ such that $n_1 + k \in I$ and for all a, b in I with $a \neq b, \mid a - b \mid \geq k$. Since the greatest common divisor of I is 1, there exists $n \in I$ such that k does not divide n. Thus $n = mk + r$ for some $m \in \mathbf{N}_0, 0 < r < k$. Since $I + I \subset I, (m+1)(n_1 + k)$ and $n + (m+1)n_1$ are in I. Their difference is $k - r < k$, a contradiction. Thus the claim holds. Now take $n_0 = n_1^2$. For $n \geq n_1^2, n - n_1^2 = mn_1 + r$ for some $m \in \mathbf{N}_0, 0 \leq r < n_1$. Thus $n = n_1^2 + mn_1 + r = (n_1+1)r + (n_1 - r + m)n_1 \in I$. □

Proof of Theorem 5.3.3 (a) Let $s \in S$ and $I = \{n \geq 1 \mid p^n(s,s) > 0\}$. Then the greatest common divisor of I is 1 and $I + I \subset I$. So there exists an $n_1 \in \mathbf{N}_0$ such that $n \in I$ for $n \geq n_1$, i.e., $p^n(s,s) > 0$ for all $n \geq n_1$. Take n_2, n_3 in $\mathbf{N}_0$ such that $p^{n_2}(i,s) > 0$ and $p^{n_3}(s,j) > 0$. Then for $n_0 = n_1 + n_2 + n_3$ (which depends on i,j), we have $p^n(i,j) > 0$ for all $n \geq n_0$. Consider an $S \times S$-valued Markov chain $\{(X_n, Y_n)\}$ with the transition matrix $\overline{P}$ given by $[[\overline{p}((i,j),(k,l))]], (i,j),(k,l) \in S \times S$, where

$$\overline{p}((i,j),(k,l)) = p(i,k)p(j,l).$$

From the foregoing, it follows that this chain is irreducible and aperiodic. It is trivial to verify that $\overline{\pi} \in \boldsymbol{P}(S \times S)$, defined by $\overline{\pi}((i,j)) = \pi(i)\pi(j)$ for $i, j \in S$, is a stationary distribution for $\{(X_n, Y_n)\}$. Thus $\{(X_n, Y_n)\}$ is positive recurrent. Let $\tau = min\{n \geq 1 \mid X_n = Y_n\}$. Since $\tau \leq \tau_{(i,i)}$ for any $i \in S$ and $(i,i) \in (S \times S)_P$, we have $\tau < \infty$ a.s. By the SMP,

$$P(X_n = j, \tau \leq n) = P(Y_n = j, \tau \leq n)$$

for all $j \in S$ and any law of (X_0, Y_0). Hence

$$\begin{aligned} P(X_n = j) &= P(X_n = j, \tau \leq n) + P(X_n = j, \tau > n) \\ &= P(Y_n = j, \tau \leq n) + P(X_n = j, \tau > n) \\ &\leq P(Y_n = j) + P(\tau > n). \end{aligned}$$

By a symmetrical argument,

$$P(Y_n = j) \leq P(X_n = j) + P(\tau > n).$$

Thus

$$| P(X_n = j) - P(Y_n = j) | \leq P(\tau > n) \to 0$$

as $n \to \infty$. Take $P(X_0 = i) = 1, P(Y_0 = k) = \pi(k)$ for $k \in S$ to conclude the proof.

(b) First we extend (a) slightly. Suppose we drop the assumption of irreducibility and assume that $S_P \neq \phi$. If $C \subset S_P$ is closed and irreducible, then there is a unique stationary distribution π supported on C. By considering the Markov chain restricted to C, we find that $p^n(i,j) \to \pi(j) = T_j^{-1}$ for all $(i,j) \in C \times C$. Coming back to (b), let $Y_n = X_{nd}, n \geq 1$. It is easily verified that $\{Y_n\}$ is an aperiodic Markov chain with transition matrix $Q = P^d$, written as $[[q(i,j)]]$. Write $Q^n = [[q^n(i,j)]], n \geq 1$. Let $\overline{\tau}_j = min\{n \geq 1 \mid Y_n = j\}$. Then $E_j[\overline{\tau}_j] = d^{-1}E_j[\tau_j] = d^{-1}T_j < \infty$. Hence j is positive recurrent for $\{Y_n\}$. Letting $C = \{i \in S \mid i \to j \text{ for } \{Y_n\}\}$, the foregoing discussion implies that $q^m(i,j) = p^{md}(i,j) \to dT_j^{-1} = d\pi(j)$.

Fix (i,j) and let $r_1 = min\{n \geq 1 \mid p^n(i,j) > 0\}$. Then $p^{r_1}(i,j) > 0$. If $p^m(j,i) > 0$ for some m, then $p^{r_1+m}(j,j) > 0$. More generally, for any n satisfying $p^n(i,j) > 0, p^{n+m}(j,j) > 0$. Hence d divides $m + r_1$ and $m + n$ and therefore, $n - r_1$. Write $r_1 = m_1 d + r$ with $0 \leq r < d, m_1 \in \mathbf{N}_0$. Then $n = md + r$ for some $m \in \mathbf{N}_0$. Since by the SMP,

$$p^n(i,j) = \sum_{k=1}^{n} P_i(\tau_j = k)p^{n-k}(j,j), n \geq 1,$$

we have

$$\begin{aligned} p^{md+r}(i,j) &= \sum_{k=0}^{m} P_i(\tau_j = kd + r)p^{(m-k)d}(j,j) \\ &= \sum_{k=0}^{\infty} P_i(\tau_j = kd + r)p^{(m-k)d}(j,j)I\{k \leq m\}. \end{aligned}$$

As $m \to \infty$,

$$p^{(m-k)d}(j,j) \to d\pi(j).$$

The dominated convergence theorem now leads to

$$p^{md+r}(i,j) \to d\pi(j)P_i(\tau_j < \infty) = d\pi(j).$$

This completes the proof. □

Theorem 5.3.4 *Let $\{X_n\}$ be positive recurrent and irreducible with π its invariant probability measure. Then π is given by*

$$\int f d\pi = E[\sum_{m=0}^{\tau_1 - 1} f(X_m)/X_0 = 1]/E[\tau_1/X_0 = 1] \tag{5.5}$$

for $f \in C_b(S)$.

Proof Let $X_0 = 1$ and $\{\sigma_i\}$ the successive return times to state 1 with $\sigma_0 = 0$. By SMP, the law of $(X_{\sigma_n}, X_{\sigma_n+1}, ...)$ coincides with its conditional law given $\mathcal{F}_{\sigma_n}$, and in turn with the law of $(X_0, X_1, ...)$ for all $n \geq 0$. It follows that for any $f \in C_b(S)$,

$$\sum_{m=\sigma_n}^{\sigma_{n+1}-1} f(X_m), n \geq 0,$$

are i.i.d. Now

$$\frac{\left(\sum_{m=0}^{\sigma_n - 1} f(X_n)\right)}{\sigma_n} = \frac{\left(\sum_{i=0}^{n-1} \left(\sum_{m=\sigma_i}^{\sigma_{i+1}-1} f(X_m)\right)\right)/n}{\left(\sum_{i=1}^{n-1} (\sigma_{i+1} - \sigma_i)\right)/n}.$$

The left-hand side a.s. tends to the left hand side of (5.5) by Theorem 5.3.2. The right-hand side in turn converges a.s. to the right-hand side of (5.5) by the strong law of large numbers. □

A sufficient condition for $\pi \in \boldsymbol{P}(S)$ to be invariant under P is given by the following:

Theorem 5.3.5 *For π as above, suppose there exists another transition matrix $Q = [[q(i,j)]]$ on S such that*

$$\pi(i)p(i,j) = \pi(j)q(j,i), \ i,j \in S.$$

Then π is invariant under P.

Proof The proof is easy (Exercise 5.5). □

This condition is extremely useful in queueing theory and interacting particle systems. We shall use it below to give an explicit expression for the invariant probability measure of a finite irreducible positive recurrent chain $\{X_n\}$.

Let $|S| = N$ and (S, E) the digraph associated with the chain, E being the set of directed edges therein. An arborescence $A \subset E$ is a set having at most one edge leaving each node, containing no cycles and maximal with respect to these two properties. It is clear that each arborescence has exactly one node for which the number of edges leaving it is zero (Exercise

5.6). Call it the root of the arborescence. Let H (resp. H_j) denote the set of arborescences (resp. arborescences with root at j), $j \in S$. The weight of an arborescence is the product of its edge weights.

Let $\| H \|$ (resp. $\| H_j \|$) denote the sum of the weights of the arborescences in H (resp. H_j).

Theorem 5.3.6 *(The Markov chain tree theorem) The invariant probability measure π for the above chain is given by $\pi(i) = \| H_i \| / \| H \|$, $i \in S$.*

Proof (Anantharam and Tsoucas) Consider transition matrices $Q = [[q(a,b)]]$ and $\overline{Q} = [[\overline{q}(a,b)]]$ on H defined as follows: $q(a,b) = p(i,j)$ if i,j are the roots of a,b respectively and b is obtained from a by adding to it a directed edge from i to j and deleting the unique edge out of j that breaks the unique directed loop created thereby. On the other hand, $\overline{q}(b,a) = p(j,k)$ if j is the root of b and a is obtained from b by adding to b some directed edge (j,k) and deleting the unique edge entering j which breaks the unique directed loop created thereby. Letting $w(a) =$ the weight of a, one checks that $\hat{\pi}(a) = w(a)/ \| H \|, a \in H$, satisfies

$$\hat{\pi}(a)Q(a,b) = \hat{\pi}(b)\overline{Q}(b,a), \ a,b \in H.$$

(Exercise 5.7) By the preceding theorem, one can construct a stationary H-valued Markov chain $\{Y_n\}$ with transition matrix Q and initial law $\hat{\pi}$. Let $f : H \to S$ be the map that maps $a \in H$ to its root. Then $X_n = f(Y_n), n \geq 0$, is seen to be a stationary chain on S with transition matrix P and the law of $X_n = \pi$ (defined as in the statement of the theorem), for $n \geq 0$. The claim follows. □

5.4 Transient and Null Recurrent Chains

In this section, we prove the analogs of some of the results of the preceding section for transient and null recurrent chains.

Theorem 5.4.1 *Let $\{X_n\}$ be irreducible and either transient or null recurrent. Then $p^n(i,j) \to 0$ as $n \to \infty$ for all $i,j \in S$.*

Proof The transient case follows from the fact that $G(i,j) < \infty$. For the null recurrent case, we may assume without any loss of generality that the chain is aperiodic. Let P^* be a limit point of $\{P^n\}$ in $[0,1]^{\infty \times \infty}$ with $P^{n(k)} \to P^*$ termwise. Argue as in part (a) of Theorem 5.3.3 to conclude that

$$| p^n(i,j) - p^n(k,j) | \to 0 \text{ as n } \to \infty, i,j,k \in S.$$

Hence all rows of P^* are identical, implying $PP^* = P^*$. By the dominated convergence theorem,

$$P^{(n(k)+1)} = PP^{n(k)} \to PP^* = P^*.$$

By Fatou's lemma,

$$P^*P = (\lim_{k\to\infty} P^{n(k)})P \le \lim_{k\to\infty} (P^{n(k)}P) = P^*,$$

where the inequality is termwise. If μ is any row of P^*, $\mu(i) \in [0,1]$ for all i and $\mu_0 = \sum_i \mu(i) \le 1$ by Fatou's lemma. If $\mu_0 > 0$, let $\overline{\mu}(i) = \mu_0^{-1}\mu(i), i \in S$. Then $\overline{\mu} \in \boldsymbol{P}(S)$ and $\overline{\mu}P \le \overline{\mu}$. Both sides add to 1. Thus equality must hold, implying that $\overline{\mu}$ is a stationary distribution for $\{X_n\}$. This contradicts the null recurrence of $\{X_n\}$, proving the claim. □

Now consider an irreducible null recurrent chain $\{X_n\}$. Let $X_0 = 1, \tau = min\{n \ge 1 \mid X_n = 1\}$ and

$$\pi(i) = E[\sum_{n=1}^{\tau} I\{X_n = i\}/X_0 = 1], i \in S.$$

Note that $M_n, n \ge 1$, defined by

$$M_n = \sum_{m=1}^{n} (I\{X_m = i\} - \sum_{j\in S} p(j,i)I\{X_{m-1} = j\}), n \ge 1,$$

is a zero-mean martingale with respect to $\sigma(X_i, i \le n), n \ge 1$, with bounded increments. A straightforward application of the optional sampling theorem to $\{M_n\}$ shows that $\pi(i) = \sum_j p(j,i)\pi(j), i \in S$ (Exercise 5.8). Since the chain is irreducible, it then follows that either $\pi(i) = \infty$ for all i or $\pi(i) < \infty$ for all i. Since $\pi(1) = 1$, the latter must hold. A similar argument shows that $\pi(i) > 0$ for all i. Thus $\pi = [\pi(1), \pi(2), ...]$ is a positive measure invariant under P. Obviously, π is not a finite measure, otherwise null recurrence of $\{X_n\}$ is contradicted.

Theorem 5.4.2 *For $i, j \in S$ and arbitrary initial law,*

$$\lim_{n\to\infty} \frac{\sum_{m=0}^{n} I\{X_m = i\}}{\sum_{m=0}^{n} I\{X_m = j\}} = \frac{\pi(i)}{\pi(j)} \quad a.s.$$

Proof Without loss of generality, let $X_0 = 1$. Let $\tau_0 = 0$ and $\{\tau_n\}$ the successive return times to 1. Then the left hand side above (Exercise 5.9) equals

$$\lim_{n\to\infty} \left[\frac{1}{n}\sum_{m=1}^{n}\left(\sum_{k=\tau_{m-1}}^{\tau_m - 1} I\{X_k = i\}\right)\right] \Big/ \left[\frac{1}{n}\sum_{m=1}^{n}\left(\sum_{k=\tau_{m-1}}^{\tau_m - 1} I\{X_k = j\}\right)\right].$$

The claim now follows from the strong law of large numbers as in Theorem 5.3.4. □

The next result shows that π is unique up to a scalar multiple.

Theorem 5.4.3 *Let ν be a nonnegative measure on S invariant under P. Then ν is a scalar multiple of π.*

Proof Let $A \subset S$ be a finite set. Define $Q = [[q(i,j)]], i,j \in A$, by

$$q(i,j) = \sum_{m=1}^{\infty} P(X_m = j, X_{m-1} \notin A, ..., X_1 \notin A / X_0 = i)$$

for $i,j \in A$. Then Q is the transition matrix for the A-valued irreducible Markov chain $\{Y_n\}$ given by: $Y_n = X_{T_n}, n \geq 1$, where $\{T_n\}$ are the successive times at which $\{X_m\}$ is in A (Exercise 5.10). Let ν_A, π_A be the restrictions of ν, π respectively to A. Since $\nu P = \nu$ (where ν is being written as a row vector),

$$\sum_{i \in A} \nu(i) p(i,j) + \sum_{i \in S \setminus A} \nu(i) p(i,j) = \nu(j), j \in A,$$

leading to

$$\sum_{i \in A} \nu(i) p(i,j) + \sum_{i \in A} \nu(i) \sum_{k \in S \setminus A} p(i,k) p(k,j) + \sum_{i \in S \setminus A} \sum_{k \in S \setminus A} \nu(i) p(i,k) p(k,j) = \nu(j).$$

Thus

$$\sum_{i \in A} \nu(i) p(i,j) + \sum_{i \in A} \nu(i) \sum_{k \in S \setminus A} p(i,k) p(k,j) \leq \nu(j).$$

Iterating, one gets (Exercise 5.11)

$$\nu_A Q \leq \nu_A.$$

Summing over both sides, we get the same number, viz., $\sum_{i \in A} \nu(i)$. So equality must prevail, implying that ν_A must be invariant under Q. The same argument also applies to π_A. Since $\{Y_n\}$ is an irreducible Markov chain on the finite state space A, it has a unique invariant probability measure and both π_A, ν_A must be scalar multiples thereof, hence of each other. Since A was arbitrary, the claim follows. □

Note that a transient chain may also have an invariant measure. Consider, for example, the chain $\{X_n\}$ on $S = \{0, \pm 1, \pm 2, ...\}$ with $X_0 = 0$ and transition probabilities $p(i, i+1) = 2/3 = 1 - p(i, i-1), i \in S$. Then $X_n = \sum_{i=1}^{n} Y_i, n \geq 1$, with $\{Y_n\}$ i.i.d., $P(Y_n = 1) = 2/3 = 1 - P(Y_n = -1), n \geq 1$.

It is easily seen that $X_n \to \infty$ a.s. and hence is transient (Exercise 5.12). However, $\pi = [\pi(1), \pi(2), ...]$ given by $\pi(i) = 1, i \in S$, is invariant under P. In absence of a counterpart of Theorem 5.4.2, the existence of an invariant measure for a transient chain is not of much use probabilistically (See, however, [45].).

In conclusion, it should be remarked that what we have been studying so far are the time-homogeneous Markov chains (also called Markov chains with stationary transitions) for which the probability of a transition from $i \in S$ to $j \in S$ is given by a fixed number $p(i, j)$ regardless of when the transition takes place. More generally, one can consider "time-inhomogeneous Markov chains" whose transitions are governed by a sequence of stochastic matrices $P_n = [[p_n(i, j)]], i, j \in S$, rather than by a single matrix P. Thus $\{P_n\}$ satisfy

$$p_n(i,j) \in [0,1], \ \sum_j p_n(i,j) = 1, \ n \geq 0, \ i,j \in S,$$

and (c) in the definition of a Markov chain given in Section 5.1 becomes

$$P(X_{n+1} \in A/\mathcal{F}_n) = P(X_{n+1} \in A/X_n) = \sum_{j \in A} p_n(X_n, j), \ n \geq 0.$$

SMP follows as before, but much of the rest of the theory developed above does not carry over to the time-inhomogeneous case. See [25] for some results on time-inhomogeneous Markov chains. For further reading, see [1, 21, 27, 35, 36, 50].

5.5 Additional Exercises

(5.13) Let $\{X_n\}$ be an irreducible Markov chain on $S_n = \{1, 2, ..., N\}$ with a transition matrix $P = [[p(i, j)]]$ which is doubly stochastic, i.e., satisfies $\sum_j p(i, j) = \sum_i p(i, j) = 1$. What is its invariant probability measure?

(5.14) Let $\{X_n\}$ be an irreducible positive recurrent Markov chain on $S = \{1, 2, ...\}$ and $\tau = min\{n \geq 1 \mid X_n = 1\}$. Show that

$$E[\tau/X_0 = i] < \infty, \forall i \in S.$$

(5.15) Let S, τ be as above. (a) Give an example of a Markov chain $\{X_n\}$ on S such that $E[\tau/X_0 = i] < \infty$ for all $i \neq 1$, but the chain is not positive recurrent. (b) Give an example of a Markov chain $\{X_n\}$ on S such that $E[\tau/X_0 = 1] < \infty$, but $E[\tau^2/X_0 = 1] = \infty$.

(5.16) Let $X_0 = 0, X_n = Y_1 + ... + Y_n, n \geq 1$, where $\{Y_n\}$ are i.i.d. with $P(Y_1 = 1) = P(Y_1 = -1) = 1/2$. (This is the "symmetric random walk".) (a) Show that $\{X_n\}$ is an irreducible null recurrent chain on $\overline{S} = \{0, \pm 1, \pm 2, ...\}$. (b) Let N, M be integers $\geq 1, \tau_N = min\{n \geq 1 \mid X_n = N], \tau_M = min\{n \geq 1 \mid X_n = -M\}$. Calculate $P(\tau_N < \tau_M)$. (c) Let $f : \overline{S} \to \boldsymbol{R}$ be a bounded function such that $(f(X_n), \sigma(X_i, i \leq n)), n \geq 0$, is a martingale. Show that f must be a constant function.

(5.17) For S as above, let $\{X_n\}$ be an irreducible chain on S and A a proper finite subset of S. Let $\tau = min\{n \geq 0 \mid X_n \notin A\}$. Show that there exist $K > 0$ and $a \in (0,1)$ such that $P(\tau > n/X_0 = i) \leq Ka^n, n \geq 1, i \in A$. (In particular, this shows that $E[\tau] < \infty$.)

(5.18) In the above set up, let $f, g : S \to \boldsymbol{R}$ be bounded. Let $\partial A = \{i \in S \backslash A \mid p(j,i) > 0$ for some $j \in A\}$ and let $\overline{A} = A \bigcup \partial A$. Show that $V(i) = E[\sum_{n=0}^{\tau-1} f(X_n) + g(X_\tau)/X_0 = i], i \in \overline{A}$, is the unique solution to the system of equations

$$\begin{array}{lll} V(i) & = f(i) + \sum_j p(i,j)V(j), & i \in A, \\ V(i) & = g(i), & i \in \partial A. \end{array}$$

(5.19) In the above set up, let $k : S \to \boldsymbol{R}$ be bounded and $b \in (0,1)$. Show that $U(i) = E[\sum_{n=0}^{\infty} b^n k(X_n)/X_0 = i], i \in S$, is the unique bounded solution to the system of equations

$$U(i) = k(i) + \beta \sum_j p(i,j)U(j),\ i \in S.$$

If k is not bounded, but is nonnegative and $U(i)$ is finite for all i, show that $U(.)$ is the least solution of the above system, that is, any other solution thereof dominates $U(.)$ termwise.

(5.20) On $\overline{S}$, consider the chain $\{X_n\}$ with the transition probabilities

$$\begin{aligned} p(i,j) &= \frac{1}{2}exp(-(G(j) - G(i))^+/T), j = i \pm 1, \\ p(i,i) &= 1 - \sum_{j=i\pm 1} p(i,j), \end{aligned}$$

for some $G : \overline{S} \to \boldsymbol{R}^+, T > 0$ satisfying $\sum_j exp(-G(j)/T) < \infty, i,j \in \overline{S}$. Show that it is irreducible positive recurrent with the unique invariant probability measure $\pi = [\pi(0), \pi(\pm 1), \pi(\pm 2), ...]$ given by $\pi(i) = exp(-G(i)/T)/[\sum_j exp(-G(j)/T)], i \in \overline{S}$. (This is called the "Gibbs distribution with potential G").

(5.21) (a) Let $\{X_n\}$ be an irreducible chain on S above. Suppose there exists a finite $A \subset S$ and a $w : S \to \boldsymbol{R}^+$ such that $w(i) \to \infty$ as $i \to \infty$ and $\sum_j p(i,j)w(j) \leq w(i)$ for $i \notin A$. Show that $\{X_n\}$ is recurrent. (b) Suppose in addition that for each $i \in S, p(i,j) > 0$ for at most finitely many j and moreover,

$$\sum_j p(i,j)w(j) \leq w(i) - \epsilon$$

for $i \notin A$ and an $\epsilon > 0$ independent of i. Show that $\{X_n\}$ is positive recurrent.

(5.22) In the set up of Theorem 5.3.5, let $\{X_n\}, \{Y_n\}$ be chains with initial law π as in the theorem and transition matrices P, Q respectively. Show that for any $m, n \geq 0$, the laws of $(X_n, X_{n+1}, ..., X_{n+m})$ and $(Y_{n+m}, Y_{n+m-1}, ..., Y_n)$ agree. (Remark: $\{Y_n\}$ is called the time-reversal of $\{X_n\}$. In particular, if $P \doteq Q$, then $\{X_n\}$ is said to be reversible.)

(5.23) If $\{X_n, 0 \leq n \leq N\}$ is a Markov chain, show that $Y_n = X_{N-n}, 0 \leq n \leq N$, is also a Markov chain, but not necessarily a time-homogeneous one.

(5.24) (a) Let $\{X_n, n = 0, \pm 1, \pm 2, \ldots\}$ be a stationary Markov chain with transition matrix $[[p(i,j)]]$ and invariant probability measure π. Let $Y_n = -X_n, \forall n$. Show that $\{Y_n\}$ is a stationary Markov chain with transition matrix $[[q(i,j)]]$ given by

$$q(i,j) = \pi(j)p(j,i)/\pi(i).$$

(This is the converse of Theorem 5.3.5.)

(b) Show that a chain with transition probabilities $\frac{1}{2}(p(i,j) + q(i,j))$ will be reversible with invariant probability measure π.

(5.25) Let (V, E) be a connected undirected graph with a finite set of vertices V and a set of edges $E \subset V \times V$. Assign to each edge (i,j) a weight $W_{ij} = W_{ji} > 0$ and set $W_i = \sum_j W_{ij}$. Show that the Markov chain on V with transition probabilities $p(i,j) = W_{ij}/W_i$ is reversible and find its invariant probability measure. Conversely, show that every irreducible reversible Markov chain on V can be obtained in this manner.

(5.26) Let $\{X_n\}$ be an irreducible positive recurrent chain on S with transition matrix P and unique invariant probability measure π. Let ν be a finite signed measure on S which is invariant under P. Show that ν must be a scalar multiple of π.

(5.27) Let $\{X_n\}$ be an S-valued random process with $X_0 = i_0$ (say) and P a stochastic matrix on S. Show that $\{X_n\}$ is a Markov chain with transition matrix P if and only if for all bounded $f : S \to \boldsymbol{R}$, $(\sum_{m=1}^{n}(f(X_m) - \sum_j p(X_{m-1}, j)f(j)), \sigma(X_m, m \leq n)), n \geq 1$, is a zero mean martingale.

6

Foundations of Continuous-Time Processes

6.1 Introduction

This chapter introduces some basics of continuous time stochastic processes, mainly those concerning their construction and sample path properties. A detailed study of these processes is a vast enterprise and is well beyond the scope of this book. For simplicity, we shall restrict to real-valued processes on the time interval $[0, 1]$, that is, a family $\{X_t\}$ of real random variables on some probability space, indexed by the time variable $t \in [0, 1]$. Much of what we do below extends to arbitrary time intervals and more general (e.g. Polish space-valued) processes with minor modifications.

A process $\{X_t, t \in [0, 1]\}$ as above will in general be prescribed through its finite dimensional distributions (also called "marginals"), that is, the laws of $[X_{t(1)}, ..., X_{t(n)}]$ for every finite subset $\{t(1), ..., t(n)\}$ of $[0, 1]$. These obviously have to be consistent. Given a consistent family of finite dimensional marginals, one can construct on $(\Omega, \mathcal{F}), \Omega = \boldsymbol{R}^{[0,1]}, \mathcal{F} =$ its product σ-field, a unique probability measure P such that the process $\{X_t, t \in [0, 1]\}$ defined by $X_t(\omega) = \omega(t)$ for $\omega = \omega(.) = \Pi_t \omega(t) \in \Omega$, has the desired finite dimensional marginals. All this is straightforward from the Kolmogorov extension theorem (Theorem 1.1.2). This construction, however, has two problems. The first of these was already remarked upon in the discussion following Theorem 1.1.2, viz., that the only subsets of Ω that are in $\mathcal{F}$ are those that can be specified in terms of countably many coordinates.

Thus important sets such as $\{\omega(.) \mid \sup_{t\in[0,1]} \omega(t) \geq 1\}$, $\{\omega(.) \mid t \to \omega(t)$ is continuous at $t = t_0\}$ are not necessarily in $\mathcal{F}$.

The second problem arises from the fact that one wants in general to view $\{X_t, t \in [0,1]\}$, not as a collection of random variables, but as a random function $t \to X_t(\omega)$ (called the "sample path" at ω). One then requires that as a function of t, $\{X_t, t \in [0,1]\}$, should have some regularity properties, such as measurability, continuity, etc., depending on the situation. These do not come free. What's worse, the finite dimensional distributions do not completely specify the sample paths. Before we look further into this issue, it will help to define certain equivalence relations on stochastic processes as follows: (We assume henceforth that the underlying probability space $(\Omega, \mathcal{F}, P)$ is complete, that is, $\mathcal{F}$ contains all subsets of sets of zero "P-measure").

Definition 6.1.1 *Two processes $X_t, Y_t, t \in [0,1]$, defined on a common probability space $(\Omega, \mathcal{F}, P)$ are said to be indistinguishable if the set $\{\omega \mid X_t(\omega) \neq Y_t(\omega)$ for some $t\}$ is contained in some set of $\mathcal{F}$ having zero P-measure. They are said to be versions of each other if $P(X_t = Y_t) = 1$ for all t.*

It is clear that indistinguishable processes will be versions of each other. The converse need not be true, as the following example shows: Let $\Omega = [0,1]$ with $\mathcal{F}$ = its Borel σ-field, completed with respect to P = the Lebesgue measure. Let $X_t(\omega) = I\{\omega = t\}, Y_t(\omega) = 0,\ t, \omega \in [0,1]$. Then $\{X_t\}, \{Y_t\}$ are only versions of each other, but are not indistinguishable (Exercise 6.1). This example also shows that though two indistinguishable processes will necessarily have a.s. indistinguishable sample paths, two processes that are versions of each other need not: $t \to X_t$ has discontinuous sample paths whereas $t \to Y_t$ has constant sample paths. It is then tempting to work with equivalence classes of indistinguishable processes, but this luxury is not permitted by the fact that our specification of the process is merely in terms of its finite dimensional marginals and two processes that are merely versions of each other will have the same finite dimensional marginals. This prompts us to look at equivalence classes of processes that are versions of each other and seek a representative of such an equivalence class that has the desired sample path properties, ignoring its versions that don't.

To work around the aforementioned difficulties, Doob introduced the concepts of measurable and separable versions which we study in the next section. Often one is interested in more than just measurable versions, such as continuous or *cadlag* versions (Here, "*cadlag*" stands for "continue a droite, limites à gauche", French for right-continuous with left limits.)

If these are available, one may view $\omega \to (t \to X_t(\omega))$ as a random variable taking values in a suitable function space, appropriately topologized and endowed with the resultant Borel σ-field. These issues are studied in sections 6.3 and 6.4. Section 6.5 lists some important classes of stochastic processes. Some useful references for this chapter are [3, 11, 15, 22, 38, 49].

6.2 Separability and Measurability

Let $\overline{\boldsymbol{R}} = [-\infty, \infty]$ denote the two point compactification of $\boldsymbol{R}$. For technical reasons that will become apparent later, the next definition involves $\overline{\boldsymbol{R}}$-valued processes rather than real-valued ones.

Definition 6.2.1 *An $\overline{\boldsymbol{R}}$-valued process $X_t(\omega), t \in [0,1]$, defined on a probability space $(\Omega, \mathcal{F}, P)$ is said to be separable with separating set $I \subset [0,1]$ if I is countable dense in $[0,1]$ and there exists $N \in \mathcal{F}$ with $P(N) = 0$ such that for any open $G \subset [0,1]$ and closed $F \subset \overline{\boldsymbol{R}}$,*

$$\{\omega \mid X_t(\omega) \in F, t \in G\} \Delta \{\omega \mid X_t(\omega) \in F, t \in G \cap I\} \subset N.$$

Before exploring the implications of this definition, we give two equivalent definitions. Let $I \subset [0,1]$ be countable dense. For each fixed $\omega \in \Omega$ and open $G \subset [0,1]$, define

$$A(G, \omega) = \overline{\{X_t(\omega) \mid t \in G \cap I\}} \subset \overline{\boldsymbol{R}},$$

where the closure is in $\overline{\boldsymbol{R}}$. Set $A(t, \omega) = \bigcap A(G, \omega)$ where the intersection is over all open $G \subset [0,1]$ containing t. This set is always nonempty (Exercise 6.2).

Lemma 6.2.1 *The following are equivalent:*

(i) $X_t, t \in [0,1]$, *is separable with separating set* I.

(ii) $X_t(\omega) \in A(t, \omega)$ *for all* t *and all* ω *outside a set* $N \in \mathcal{F}$ *with* $P(N) = 0$.

(iii) The graph of $t : I \to X_t(\omega)$ *is dense in the graph of* $t : [0,1] \to X_t(\omega)$ *for all* ω *outside a set* $N \in \mathcal{F}$ *with* $P(N) = 0$.

Proof That (i) implies (ii) follows easily from the definition of separability (Exercise 6.3). Conversely, let (ii) hold. Then for any $\omega \notin N$, closed $F \subset \overline{\boldsymbol{R}}$ and open $S \subset [0,1]$, the following holds: If $X_t(\omega) \in F$ for $t \in I \bigcap S$, then $A(t,\omega) \subset A(S,\omega) \subset F$ for $t \in S$, implying $X_t(\omega) \in F$ for $t \in S$. Thus (i) holds. Equivalence of (ii) and (iii) is easily established (Exercise 6.4). □

To appreciate the importance of separability, recall the probability space $(\Omega, \mathcal{F}, P)$ with $\Omega = \boldsymbol{R}^{[0,1]}$, $\mathcal{F}$ = its product σ-field completed with respect to P, and the process $X_t, t \in [0,1]$, defined on $(\Omega, \mathcal{F}, P)$ by $X_t(\omega) = \omega(t), t \in [0,1]$. If $\{X_t\}$ is separable with I, N as above, then

$$\{\omega \mid \sup_t X_t(\omega) \geq 1\} \Delta \{\omega \mid \sup_{t\in I} X_t(\omega) \geq 1\} \subset N$$

and

$$\{\omega \mid t \to X_t(\omega) \textit{ is continuous at } t_0\} \Delta \{\omega \mid \lim_{n\to\infty} \sup_{|t-t_0|<1/n, t\in I} X_t$$
$$= \lim_{n\to\infty} \inf_{|t-t_0|<1/n, t\in I} X_t = X_{t_0}\} \subset N.$$

The second set of each pair on the left is in $\mathcal{F}$ and hence, by completeness of $(\Omega, \mathcal{F}, P)$, so is the first. Contrast this with the remarks regarding the latter made at the beginning of the preceding section.

The main result concerning separability is that a real-valued process always has a separable version if we allow $\pm\infty$ as possible values. We shall prove this below following two technical lemmas. Let $X_t, t \in [0,1]$, be a real-valued process on some complete probability space $(\Omega, \mathcal{F}, P)$.

Lemma 6.2.2 *Let $B \subset \overline{\boldsymbol{R}}$ be any Borel set. Then there exists an at most countable set $\{t(n)\} \subset [0,1]$ such that*

$$P(X_{t(k)} \in B, k \geq 1, X_t(\omega) \notin B) = 0, t \in [0,1].$$

Proof Pick the $\{t(n)\}$ recursively as follows: Let $t(1)$ be arbitrary. Given $t(1), ..., t(k)$, let

$$m_k = \sup_{t\in[0,1]} P(X_{t(i)} \in B, 1 \leq i \leq k, X_t \notin B).$$

If $m_k = 0$, we are done. If not, pick $t(k+1)$ such that

$$P(X_{t(i)} \in B, 1 \leq i \leq k, X_{t(k+1)} \notin B) \geq m_k/2.$$

The sets $L_k = \{X_{t(i)} \in B_i, 1 \leq i \leq k, X_{t(k+1)} \notin B\}, k \geq 1$, are mutually disjoint. Thus

$$1 \geq \sum_k P(L_k) \geq \frac{1}{2} \sum_k m_k,$$

implying $m_k \to 0$ as $k \to \infty$. Thus

$$P(X_{t(k)} \in B, k \geq 1, X_t \notin B) \leq \lim_{k\to\infty} m_k = 0.$$ □

Lemma 6.2.3 *Let Q_0 be a countable collection of Borel subsets of $\overline{\boldsymbol{R}}$ and Q the collection of all possible intersections of subcollections of sets belonging to Q_0. Then there exists an at most countable set $\{t(i), i \geq 1\} \subset [0,1]$ such that for each $t \in [0,1]$, there exists a set $N_t \in \mathcal{F}$ with $P(N_t) = 0$ and $\{X_{t(k)} \in B,\ k \geq 1,\ X_t \notin B\} \subset N_t$ for all $B \in Q$.*

Proof For each fixed $B \in Q_0$, construct the at most countable set $\{t(k)\} \subset [0,1]$ as in the preceding lemma and let I denote the union of all such sets. Let $t \in [0,1]$ and

$$N_t = \bigcup_{B \in Q_0} \{X_s \in B, s \in I, X_t \notin B\}.$$

By the preceding lemma, $P(N_t) = 0$. If $C \in Q$ and $C \subset B \in Q_0$, then

$$\{X_s \in C, s \in I, X_t \notin B\} \subset \{X_s \in B, s \in I, X_t \notin B\} \subset N_t.$$

If $C = \bigcap_{k=1}^{\infty} B_k$ for $\{B_1, B_2, ...\} \subset Q_0$, then

$$\{X_s \in C, s \in I, X_t \notin C\} \subset \bigcup_{k=1}^{\infty} \{X_s \in C, s \in I, X_t \notin B_k\} \subset N_t.$$

The claim follows. □

Theorem 6.2.1 *There exists an $\overline{\boldsymbol{R}}$-valued separable process $\overline{X}_t, t \in [0,1]$ on $(\Omega, \mathcal{F}, P)$ which is a version of $X_t, t \in [0,1]$.*

Proof View $X_t, t \in [0,1]$, as an $\overline{\boldsymbol{R}}$-valued process. Let Q_0 be the collection of complements of open intervals in $\overline{\boldsymbol{R}}$ with rational centres and rational radii (*w.r.t.* any compatible metric). Let Q be as above. Then Q contains all closed subsets of $\overline{\boldsymbol{R}}$. Let $\mathcal{B}$ be the collection of open intervals in $[0,1]$ with rational centres and rational radii. Consider $X_t, t \in G$, for $G \in \mathcal{B}$. By the foregoing, there is a countable dense subset $I(G)$ of G and a set $N_t(G) \in \mathcal{F}$ such that $P(N_t(G)) = 0$ and for each $B \in Q$,

$$\{X_s \in B, s \in I(G), X_t \notin B\} \subset N_t(G).$$

Let $I = \bigcup_{G \in \mathcal{B}} I(G), N_t = \bigcup_{G \in \mathcal{B}} N_t(G)$. Define

$$\begin{aligned} \overline{X}_t(\omega) &= X_t(\omega), \text{ if either } t \in I \text{ or } \omega \notin N_t, \\ &= \text{any element of } A(t,\omega), \text{ otherwise.} \end{aligned}$$

Since $X_t(\omega) = \overline{X}_t(\omega)$ for $t \in I, A(t,\omega) = \overline{A}(t,\omega)$ for all t where

$$\overline{A}(t,\omega) = \bigcap_{t \in G, G \text{ open}} \overline{\{\overline{X}_s(\omega) \mid s \in G \cap I\}},$$

the closures on the right hand side being in $\overline{\boldsymbol{R}}$. By the above lemma and the definition of $\{\overline{X}_t\}$, it then follows that $\overline{X}_t(\omega) \in \overline{A}(t,\omega)$ for all t,ω (Exercise 6.5). By Lemma 6.2.1, $\overline{X}_t, t \in [0,1]$, is separable. Also, $\{\omega \mid X_t(\omega) \neq \overline{X}_t(\omega)\} \subset N_t, t \in [0,1]$. Thus $\{X_t\}, \{\overline{X}_t\}$ are versions of each other. This completes the proof. □

Call $X_t, t \in [0,1]$, stochastically continuous (or continuous in probability) if for all $t \in [0,1]$ and $\epsilon > 0$, $P(\mid X_t - X_s \mid > \epsilon) \to 0$ as $s \to t$. Note that this property depends only on the two dimensional marginals of the process. The following theorem covers many stochastic processes of interest:

Theorem 6.2.2 *If $\{X_t, t \in [0,1]\}$ in the preceding theorem is stochastically continuous, then any countable dense $J \subset [0,1]$ may be taken as a separating set for $\{\overline{X}_t\}$.*

Proof Let $I = \{t(1), t(2), ...\}$ be as in the above theorem and J any countable dense subset of $[0,1]$. Let $\{s(i,j), i \geq 1, j \geq 1\} \subset J$ be such that $s(i,n) \to t(i)$ for each i. Then $X_{s(i,n)} \to X_{t(i)}$ in probability for each i and therefore $X_{s(i,n(i,m))} \to X_{t(i)}$ a.s. for a subsequence $\{n(i,m)\}$ of $\{n\}, i \geq 1$. Let N be the zero probability set where either this convergence fails for some i or the graph of $t \in I \to X_t(\omega)$ is not dense in the graph of $t \in [0,1] \to \overline{X}_t(\omega)$. Then for $\omega \notin N$, the graph of $t \in I \to X_t(\omega)$ and hence the graph of $t \in J \to X_t(\omega)$ is dense in the graph of $t \in [0,1] \to \overline{X}_t(\omega)$. The claim follows by Lemma 6.2.1. □

In the above, $X_t = \overline{X}_t$ a.s. for each t and thus $\mid \overline{X}_t \mid < \infty$ a.s. for each t. This does not, however, imply that a.s., $\mid X_t \mid < \infty$ for all $t \in [0,1]$, as the following example (due to V.V. Phansalkar) shows: Let $\Omega = [0,1]$ and $\mathcal{F} =$ its Borel σ-field, completed with respect to $P =$ the Lebesgue measure. On $(\Omega, \mathcal{F}, P)$, define the process $X_t, t \in [0,1]$, by $X_t(\omega) = (t-\omega)^{-1} I\{t \neq \omega\}$. One checks that this does not have an $\boldsymbol{R}$-valued separable version, but $\overline{X}_t(\omega) = (t-\omega)^{-1} I\{t \neq \omega\} + \infty I\{t = \omega\}, t \in [0,1]$, does present an $\overline{\boldsymbol{R}}$-valued separable version (Exercise 6.6).

We shall now address the measurability issue.

Definition 6.2.2 *A process $X_t, t \in [0,1]$, defined on a probability space $(\Omega, \mathcal{F}, P)$ is said to be measurable if the map $(t,\omega) \in [0,1] \times \Omega \to X_t(\omega) \in \boldsymbol{R}$ is measurable with respect to the product σ-field.*

The following example shows that measurability cannot be taken for granted: Let $X_t, t \in [0,1]$, be independent and identically distributed random variables, uniformly distributed on $[-1,1]$. Let $\{X_t\}$ have a measurable version, which we also denote by $\{X_t\}$. Let $Y_t = \int_0^t X_s ds, t \in [0,1]$. Each Y_t is a random variable (Exercise 6.7). Also, $t \to Y_t(\omega)$ is continuous for all ω. If $P(Y_t \neq 0) = 0$ for all $t \in [0,1], Y_t = 0$ for all rationals a.s. and hence for all t a.s. by path continuity. But then $X_t = 0$ for a.e. t, a.s. and hence a.s. for all t. Thus there must exist a $t \in [0,1]$ such that $P(Y_t \neq 0) > 0$. Then $E[Y_t^2] > 0$. But

$$E[Y_t^2] = E[\int_0^t \int_0^t X_s X_y dsdy] = \int_0^t \int_0^t E[X_s X_y]dsdy$$
$$= E[X_1^2] \int_0^t \int_0^t I\{s = y\}dsdy = 0,$$

a contradiction. Thus $\{X_t\}$ cannot have a measurable version.

The following theorem covers many cases of interest. As in the preceding theorems, we view $X_t, t \in [0,1]$ as an $\overline{\boldsymbol{R}}$−valued process.

Theorem 6.2.3 *Let $X_t, t \in [0,1]$, be stochastically continuous. Then it has a separable measurable version.*

Proof In view of the preceding theorems, we may suppose that $X_t, t \in [0,1]$, is separable with separating set I. For each $n \geq 1$, cover $[0,1]$ with finitely many open sets $S(n,k), k = 1, 2, ..., m(n)$, of diameter not exceeding $1/n$. Pick $t(n,k) \in I \bigcap S(n,k)$ and let $B(n,k) = S(n,k) \setminus (\bigcup_{j=1}^{k-1} S(n,j))$ for $k = 1, ..., m(n)$. Define

$$X_t^n(\omega) = \sum_{k=1}^{m(n)} X_{t(n,k)}(\omega) I_{B(n,k)}(t), \ t \in [0,1], n \geq 1.$$

Let $\epsilon > 0$ and

$$G_{n,m}(t) = P(\mid X_t^n - X_t^m \mid > \epsilon).$$

Since $X_t^n = X_{t(n,k)}$ and $\mid t(n,k) - t \mid < \frac{1}{n}$ for $t \in B(n,k)$, stochastic continuity of $\{X_t\}$ implies that $G_{n,m}(t) \to 0$ as $n, m \to \infty$. Letting μ denote the product measure Lebesgue $\times P$ on $[0,1] \times \Omega$, we then have

$$\mu(\{(t,\omega) \mid \mid X_t^n(\omega) - X_t^m(\omega) \mid > \epsilon\}) = \int G_{n,m}(t)dt \to 0$$

as $n, m \to \infty$. Thus $\{X_t^n(\omega)\}$ is Cauchy in μ-probability and therefore converges in μ-probability. Then a subsequence thereof converges μ-a.s. to, say, $\tilde{X}_t(\omega), (t,\omega) \in [0,1] \times \Omega$. Let $N \subset [0,1] \times \Omega$ be the set of zero μ-probability where this convergence fails. It is easily seen that for $(t,\omega) \notin N, \tilde{X}_t(\omega) \in A(t,\omega)$ for $A(t,\omega)$ defined as in Lemma 6.2.1. By suitably modifying $\tilde{X}_t(\omega)$ on N, we may suppose that this holds for all (t,ω). Let $N_t = \{\omega \mid (t,\omega) \in N\}, t \in [0,1]$, and $K = \{t \mid P(N_t) > 0\}$. Then the Lebesgue measure of K must be zero. Set $\overline{X}_t(\omega) = X_t(\omega)$ for $t \in I \cup K$ and $= \tilde{X}_t(\omega)$ otherwise. From Lemma 6.2.1, it follows that $\{\overline{X}_t\}$ is separable. It is clearly measurable with respect to the product σ-field of $[0,1] \times \Omega$.

Finally, $\overline{X}_t(\omega) = X_t(\omega)$ for $t \in I \cup K$ and a.s. for $t \notin I \cup K$ by stochastic continuity (Exercise 6.8). Thus $\{\overline{X}_t\}$ is a version of $\{X_t\}$. This completes the proof. □

The requirement of stochastic continuity may be replaced by the weaker requirement of stochastic continuity at a.e. t in $[0, 1]$ by modifying the above proof slightly (Exercise 6.9). As for the need to allow $\pm\infty$ as possible values for $\{\overline{X}_t\}$, this has to be done away with, if at all, by using additional information about the finite dimensional distributions of $\{X_t\}$.

We conclude this section by mentioning a refinement of the concept of measurability that plays a major role in the so called "general theory of processes". Let $(\Omega, \mathcal{F}, P)$ be a complete probability space and $\{\mathcal{F}_t, t \geq 0\}$ an increasing right-continuous (i.e., $\mathcal{F}_t = \bigcap_{s>t} \mathcal{F}_s$ for all t) family of complete sub-σ-fields of $\mathcal{F}$. Let $\mathcal{B}_t$ denote the Borel σ-field of $[0, t], t \in [0, 1]$, and $\mathcal{G}_t =$ the product σ-field $\mathcal{B}_t \times \mathcal{F}_t$ completed with respect to Lebesgue $\times P$-measure, $t \geq 0$. A real-valued process $\{X_t, t \in [0, 1]\}$ defined on $(\Omega, \mathcal{F}, P)$ is said to be progressively measurable with respect to $\{\mathcal{F}_t\}$ if, for every t, the map $(s, \omega) \in [0, t] \times \Omega \to X_s(\omega) \in \boldsymbol{R}$ is $\mathcal{G}_t$-measurable. See [11] for more along these lines.

6.3 Continuous Versions

Let $X_t, t \in [0, 1]$, be a real-valued stochastic process defined on a complete probability space $(\Omega, \mathcal{F}, P)$. In this section, we shall study conditions under which it has a version $\{\overline{X}_t\}$ such that $t \to \overline{X}_t(\omega)$ is continuous for all ω. Let $D \subset [0, 1]$ be countable dense and for $\delta > 0, f : [0, 1] \to \boldsymbol{R}$, define the "modulus of continuity" of f on D as

$$w_D(f, \delta) = sup\{| f(t) - f(s) |:| t - s | \leq \delta,\ t, s \in D\}.$$

Lemma 6.3.1 *$\{X_t\}$ has a continuous version if and only if*

(i) $\{X_t\}$ is stochastically continuous, and

(ii) $w_D(X, \delta) \to 0$ in probability as $\delta \to 0$.

Proof Suppose $\{X_t\}$ has a continuous version $\{\overline{X}_t\}$. Then for any $t \in [0, 1], \overline{X}_s \to \overline{X}_t$ a.s. as $s \to t$ and therefore in probability. Since finite dimensional marginals of $\{X_t\}$ and $\{\overline{X}_t\}$ agree and convergence in probability depends only on two dimensional marginals, it follows that $\{X_t\}$ is stochastically continuous. Also, $w_D(\overline{X}, \delta) \to 0$ a.s. Since $\overline{X}_t = X_t, t \in D$, outside a common set of zero probability, $w_D(X, \delta) \to 0$ a.s. and hence in probability.

Conversely, let (i), (ii) hold. Since $w_D(X, \delta)$ decreases monotonically with δ, (ii) implies that $w_D(X, \delta) \to 0$ a.s. as $\delta \to 0$. Now any $f : D \to \boldsymbol{R}$ has a continuous extension $\overline{f} : [0,1] \to \boldsymbol{R}$ if and only if $w_D(f, \delta) \to 0$ as $\delta \to 0$ (Exercise 6.10). For each $\omega \notin N = \{\omega' \mid w_D(X(\omega'), \delta) \not\to 0 \text{ as } \delta \to 0\}$, let $\overline{X} : [0,1] \to \boldsymbol{R}$ denote the unique continuous extension of $X : D \to \boldsymbol{R}$. For $\omega \in N$, let $\overline{X}_t(\omega) = 0, t \in [0,1]$. Then $t \to \overline{X}_t(\omega)$ is continuous for all ω. Also, $\overline{X}_t = X_t$ a.s. for $t \in D$. For $t \notin D$, let $\{t(n)\} \subset D$ be such that $t(n) \to t$. Then $\overline{X}_{t(n)} \to \overline{X}_t$ and $X_{t(n)} = \overline{X}_{t(n)}$ a.s. together imply that $X_{t(n)} \to \overline{X}_t$ a.s. Since $X_{t(n)} \to X_t$ in probability, $X_t = \overline{X}_t$ a.s. Thus $\{X_t\}, \{\overline{X}_t\}$ are versions of each other. This completes the proof. □

Condition (ii) is not easily verifiable in practice, so one looks for easily verifiable sufficient conditions that imply (i), (ii). The following test due to Kolmogorov which involves only two dimensional marginals is extremely useful in practice.

Theorem 6.3.1 *Suppose* $\{X_t\}$ *satisfies*

$$E[\mid X_t - X_s \mid^a] \leq b \mid t - s \mid^{1+c}, t, s \in [0,1], \tag{6.1}$$

for some $a, b, c > 0$. *Then it has a continuous version.*

Proof Let $D = \{\frac{k}{2^n}, k = 0, 1, ..., 2^n, n = 0, 1, 2, ...\}$ and define

$$Z_n(\omega) = \sup_{0 \leq k \leq 2^n - 1} \mid X_{(k+1)/2^n}(\omega) - X_{k/2^n}(\omega) \mid, \; n = 1, 2, \ldots.$$

If $t, s \in [0,1]$ satisfy $\mid t - s \mid < 2^{-n}$, then we can find a k, $0 < k < 2^n$, such that one has $\mid t - k/2^n \mid, \mid s - k/2^n \mid < 2^{-n}$. If in addition, $t, s \in D, t$ must be of the form

$$t = \frac{k}{2^n} \pm \sum_{j=1}^{m} a_j 2^{-(n+j)}$$

where $a_j = 0$ or $1, 1 \leq j \leq m$, and $m \geq 1$. Then

$$\mid X_t - X_{k/2^n} \mid \leq \sum_{j=n+1}^{n+m} Z_j \leq \sum_{j=n+1}^{\infty} Z_j.$$

A similar argument applied to s in place of t leads to

$$\mid X_s - X_{k/2^n} \mid \leq \sum_{j=n+1}^{\infty} Z_j.$$

Thus

$$w_D(X, 2^{-n}) = \sup_{t,s \in D, |t-s| < 2^{-n}} \mid X_t - X_s \mid \leq 2 \sum_{j=n+1}^{\infty} Z_j.$$

By Chebyshev's inequality,

$$P(| X_t - X_s | \geq \epsilon) \leq E[| X_t - X_s |^a]/\epsilon^a, \epsilon > 0.$$

Set $h = | t - s |, \epsilon = h^r$ for some $r > 0$. Then the above inequality in conjunction with (6.1) leads to

$$P(| X_t - X_s | \geq h^r) \leq bh^{1+c-ar}.$$

Let $0 < r < c/a$ and $\delta = c - ar > 0$. Then

$$P(| X_t - X_s | \geq h^r) \leq bh^{1+\delta}$$

and

$$\begin{aligned} P(Z_n \geq 2^{-nr}) &= P(\sup_{0 \leq k < 2^n} | X_{(k+1)/2^n} - X_{k/2^n} | \geq 2^{-nr}) \\ &\leq \sum_{k=0}^{2^n-1} P(| X_{(k+1)/2^n} - X_{k/2^n} | \geq 2^{-nr}) \\ &\leq 2^n b 2^{-n(1+\delta)} = b2^{-n\delta}. \end{aligned}$$

Since

$$\sum_n 2^{-n\delta} < \infty,$$

we have

$$\sum_n P(Z_n \geq 2^{-nr}) < \infty.$$

By the Borel–Cantelli lemma, it follows that $Z_n \geq 2^{-nr}$ for at most finitely many n, a.s. Thus $\lim_{n\to\infty} \sum_{m=n+1}^{\infty} Z_m = 0$ a.s. and therefore $w_D(X, 2^{-n}) \to 0$ a.s. Since $w_D(X, \delta)$ is a nonincreasing function of δ, $w_D(X, \delta) \to 0$ a.s. as $\delta \to 0$. Since (6.1) implies stochastic continuity of $\{X_t\}$, the claim now follows from the preceding lemma. □

The above proof contains a useful estimate on the modulus of continuity, as proved below.

Corollary 6.3.1 *Let $\{X_t\}$ be a process with continuous sample paths, satisfying (6.1). Then for any $\epsilon \in (0, c/a)$,*

$$\lim_{\alpha\to 0} \alpha^{\epsilon - \frac{c}{a}} w_{[0,1]}(X, \alpha) = 0 \ a.s.$$

Proof In the notation of the preceding theorem, let $N(\omega)$ be the a.s. finite integer-valued random variable such that $Z_n(\omega) < 2^{-nr}$ for $n \geq N(\omega)$. Then arguing as above,

$$w_D(X, 2^{-n}) \leq 2 \sum_{m=n+1}^{\infty} 2^{-mr} = 2^{1-(n+1)r}/(1 - 2^{-r})$$

for $n \geq N(\omega), r \in (0, c/a)$. Thus for any $h < 2^{-N(\omega)}$,

$$w_{[0,1]}(X,h) = w_D(X,h) \leq Ch^r$$

for a suitable constant C. The claim follows. □

Suppose $\{X_t, t \in [0,1]\}$ is a process on some complete probability space $(\Omega, \mathcal{F}, P)$ satisfying conditions (i), (ii) of Lemma 6.3.1 and $\{\overline{X}_t\}$ its continuous version constructed as in the proof of the lemma. It is clear that if $\{X'_t\}$ is another continuous version of $\{X_t\}$, then $\{\overline{X}_t\}, \{X'_t\}$ are indistinguishable. The trajectory $t \to \overline{X}_t$, denoted by $\overline{X}$, may be viewed as an element of $C[0,1]$ = the Banach space of continuous functions $[0,1] \to \boldsymbol{R}$ with the supremum norm.

Theorem 6.3.2 *$\overline{X}$ is a $C[0,1]$-valued random variable, that is, the map $\omega \in \Omega \to \overline{X}(\omega) \in C[0,1]$ is $\mathcal{F}$-measurable.*

Proof This is immediate from the easily proved fact that the Borel σ-field of $C[0,1]$ is generated by the sets of the type $\{x(.) \in C[0,1] \mid x(t_i) \in A_i, 1 \leq i \leq n\}$ for $n \geq 1, t_1, ..., t_n$ rational, $A_1, \ldots, A_n \subset \boldsymbol{R}$ Borel (Exercise 6.11). □

In particular, $\{\overline{X}_t\}$ can be "canonically" realized on the probability space $(C[0,1], \mathcal{B}, \mu)$ where $\mathcal{B}$ = the Borel σ-field of $C[0,1]$ and μ = the law of $\overline{X}$ above, by setting $\overline{X}_t(\omega) = \omega(t)$ for $\omega = \omega(.) \in C[0,1]$.

6.4 *Cadlag* Versions

Recall that *cadlag* functions are those functions from $[0,1]$ to $\boldsymbol{R}$ (more generally, from any interval in $\boldsymbol{R}$ to a Polish space) that are continuous from the right and have limits from the left at each point of $[0,1]$. (They are also called "*r.c.l.l.*" for "right-continuous with left limits".) It is easy to see that these are bounded and measurable (Exercise 6.12). Also, given any $a < b$ in $\boldsymbol{R}$, the graph of a *cadlag* function $[0,1] \to \boldsymbol{R}$ crosses the strip $[a,b]$ at most finitely many times, implying in particular that it has at most countably many points of discontinuity (Exercise 6.13).

Let D be countable dense in $[0,1]$ and define

$$w'_D(f,\delta) = sup \mid f(t_3) - f(t_2) \mid \wedge \mid f(t_2) - f(t_1) \mid$$

for $f : [0,1] \to \boldsymbol{R}, \delta > 0$, the supremum being over all $t_1, t_2, t_3 \in D$ satisfying $t_1 < t_2 < t_3$ and $t_3 < t_1 + \delta$. For *cadlag* functions, $w'_D(f,\delta)$ plays a role analogous to what $w_D(f,\delta)$ did for the continuous functions. Also define $U_D(f,[a,b]) = sup\{n \mid \exists\, 0 \leq s_1 < t_1 < s_2 < t_2 < \ldots < s_n < t_n \leq 1$ such that $s_i, t_i \in D$ and $f(s_i) \leq a < b \leq f(t_i),\ 1 \leq i \leq n\}$, for $f : [0,1] \to \boldsymbol{R}$

and $a < b$ in $\boldsymbol{R}$. This is the "number of upcrossings" of $[a, b]$ by the function $f : D \to \boldsymbol{R}$.

Lemma 6.4.1 *Let $f : [0,1] \to \boldsymbol{R}$. The following are equivalent:*

(i) $\overline{f}(t) = \lim_{D \ni s \to t_+} f(s)$ exists for all t and defines a cadlag *function,*

(ii) $w'_D(f, \delta) \to 0$ as $\delta \to 0$,

(iii) f is bounded and $U_D(f, [a, b]) < \infty$ for all $a < b$ in $\boldsymbol{R}$.

(Here, $s \to t_+$ indicates a strictly decreasing limit.)

The proof is easy (Exercise 6.14). This leads to the following analog of Lemma 6.3.1, proved along similar lines (Exercise 6.15).

Lemma 6.4.2 *A real-valued process $\{X_t, t \in [0,1]\}$ has a* cadlag *version if and only if (i) it is stochastically right-continuous (i.e., $X_s \to X_t$ in probability whenever s decreases to t) and either (ii) $w'_D(X, \delta) \to 0$ in probability as $\delta \to 0$, or, (ii$'$) $\sup_{t \in D} \mid X_t \mid < \infty$ a.s. and $U_D(X, [a, b]) < \infty$ a.s. for all $a < b$ in $\boldsymbol{R}$.*

As in the case of Lemma 6.3.1, this does not provide an easily verifiable criterion for the existence of *cadlag* versions. The following extension of Theorem 6.3.1 is very useful in practice, as it involves only the three dimensional marginals of the process.

Theorem 6.4.1 *(Chentsov) If a process $\{X_t\}$ is stochastically continuous and satisfies*

$$E[\mid X_t - X_s \mid^a \mid X_s - X_u \mid^b] \leq C \mid t - u \mid^{1+d} \tag{6.2}$$

for some $a, b, C, d > 0$ and all $0 \leq u \leq s \leq t \leq 1$, then it has a cadlag *version.*

Though this result is essentially in the spirit of Kolmogorov's test (Theorem 6.3.1), its proof is much more difficult and is omitted. An interested reader may refer to [22, pp. 159–164], for details.

Just as we viewed continuous sample path processes as $C[0,1]$-valued random variables (cf. Theorem 6.3.2), we would like to view *cadlag* processes as random variables taking values in a suitable function space. The most convenient space for this purpose turns out to be $D[0,1]$ = the space of *cadlag* functions from $[0,1]$ to $\boldsymbol{R}$ with the "Skorohod topology" we describe below. Before doing so, we first observe that the topology induced by the supremum norm is not appropriate for this space. For one thing, it renders this space nonseparable, because the collection $\{f_u(.), u \in [0,1]\}$

defined by $f_u(t) = I\{t \geq u\}, t \in [0,1]$, is uncountable and any two distinct elements of it are at unit distance from each other with respect to the supremum norm. Also, one would like to have $f_u \to f_v$ whenever $u \to v$ in $[0,1]$. This is clearly impossible in any topology that leads to a convergence concept that implies pointwise convergence. Clearly, what is needed is something like "pointwise convergence after a suitable rescaling of the time axis that becomes asymptotically negligible". It is precisely this intuition that is captured by the Skorohod topology.

Let $\Lambda = \{\lambda : [0,1] \to [0,1] \mid \lambda$ is strictly increasing and onto (hence continuous)$\}$ denote the set of "scaling functions". In particular, each $\lambda \in \Lambda$ satisfies $\lambda(0) = 0$ and $\lambda(1) = 1$. For $x = x(.), y = y(.)$ in $D[0,1]$ and $\lambda \in \Lambda$, define

$$
\begin{aligned}
d(x,y) &= inf\{\epsilon > 0 \mid \exists \lambda \in \Lambda \ni \sup_t \mid \lambda(t) - t \mid \leq \epsilon \text{ and} \\
&\quad \sup_t \mid x(t) - y(\lambda(t)) \mid \leq \epsilon\}, \\
\| \lambda \| &= \sup_{s \neq t} \mid \ln[(\lambda(t) - \lambda(s))/(t-s)] \mid, \\
d_1(x,y) &= inf\{\epsilon > 0 \mid \exists \lambda \in \Lambda \ni \| \lambda \| \leq \epsilon,\ \sup_t \mid x(t) - y(\lambda(t)) \mid \leq \epsilon\}.
\end{aligned}
$$

Lemma 6.4.3 *d, d_1 define metrics on $D[0,1]$.*

Proof Let $x = x(.), y = y(.) \in D[0,1]$. Considering $\lambda(t) = t$ and using the fact that *cadlag* functions are bounded, one sees that $d(x,y)$ is always finite. Also, $d(x,y) = 0$ implies that there exist $\lambda_n \in \Lambda, n \geq 1$, such that $\lambda_n(t) \to t, y(\lambda_n(t)) \to x(t)$ uniformly in t. It follows that either $x(t) = y(t)$ or $x(t) = y(t-)$ for each t. Since $x \in D[0,1]$, the former must hold. Thus $x = y$. If $\lambda \in \Lambda, \lambda^{-1} \in \Lambda$ and vice-versa, where λ^{-1} is the inverse function. Then

$$
\begin{aligned}
\sup_t \mid \lambda(t) - t \mid &= \sup_t \mid \lambda^{-1}(t) - t \mid, \\
\sup_t \mid x(t) - y(\lambda(t)) \mid &= \sup_t \mid x(\lambda^{-1}(t)) - y(t) \mid .
\end{aligned}
$$

Thus $d(x,y) = d(y,x)$. Finally, for $\lambda_1, \lambda_2 \in \Lambda$ and $\lambda = \lambda_2 \circ \lambda_1 \in \Lambda$,

$$
\begin{aligned}
\sup_t \mid \lambda(t) - t \mid &\leq \sup_t \mid \lambda(t) - \lambda_1(t) \mid + \sup_t \mid \lambda_1(t) - t \mid \\
&= \sup_t \mid \lambda_2(t) - t \mid + \sup_t \mid \lambda_1(t) - t \mid .
\end{aligned}
$$

Similarly, for $x, y, z \in D[0,1]$,

$$
\sup_t \mid x(t) - z(\lambda(t)) \mid \leq \sup_t \mid x(t) - y(\lambda_1(t)) \mid + \sup_t \mid y(t) - z(\lambda_2(t)) \mid .
$$

It follows that $d(x,z) \leq d(x,y) + d(y,z)$. Hence d is a metric. That d_1 is a metric follows along similar lines using the relations (Exercise 6.16)

$$\| \lambda^{-1} \| = \| \lambda \|, \| \lambda_2 \circ \lambda_1 \| \leq \| \lambda_1 \| + \| \lambda_2 \| . \qquad \square$$

Lemma 6.4.4 *The metrics d, d_1 are equivalent.*

The proof relies upon the characterization of *cadlag* functions given in Lemma 6.4.5 below. For $f : [0,1] \to \boldsymbol{R}, \epsilon > 0$, define

$$\alpha_f(\epsilon) = \inf \max_{1 \leq i \leq n} (sup\{| f(t) - f(s) | \mid t, s \in [t_{i-1}, t_i)\}$$

where the infimum is over all partitions $0 = t_0 < t_1 < \ldots t_n = 1$ of [0,1] satisfying $t_{i+1} - t_i > \epsilon$ for all i.

Lemma 6.4.5 *$f : [0,1] \to \boldsymbol{R}$ is* cadlag *if and only if $\alpha_f(\epsilon) \to 0$ as $\epsilon \to 0$.*

Proof Suppose f is *cadlag*. To prove that $\alpha_f(\epsilon) \to 0$ as $\epsilon \to 0$, we need to prove that for each $\delta > 0$, there exists an $\epsilon > 0$ and a partition $\{t_i\}$ of $[0,1]$ with $t_{i+1} - t_i > \epsilon$ and $sup\{| f(t) - f(s) | \mid t, s \in [t_i, t_{i+1})\} < \delta$ for all i. Let T be the supremum of those $t_0 \in [0,1]$ for which $[0, t_0)$ can be decomposed into finitely many disjoint intervals of the type $[a,b), b > a$, satisfying $\sup\{| f(t) - f(s) | \mid t, s \in [a,b)\} < \delta$. Since $f(0) = f(0+), T > 0$. Since $f(T-)$ exists, the interval $[0,T)$ itself can be thus decomposed. If $T < 1, f(T) = f(T+)$ implies that we can find $T' > T$ in $[0,1]$ for which $[T, T')$ satisfies the condition stipulated for $[a,b)$ above. This contradicts the definition of T. Thus $T = 1$. Letting $\epsilon < min\{b - a \mid [a,b)$ is some interval in the above partition of $[0,1)\}$ completes the proof of the "only if" part. The "if" part of the claim is left as an exercise (Exercise 6.17). □

Proof of Lemma 6.4.4 For $x \in D[0,1]$, let $y \in D[0,1]$ and $\epsilon \in (0, 1/4)$ be such that $d_1(x,y) < \epsilon$. Then for some $\lambda \in \Lambda$,

$$\sup_t | x(t) - y(\lambda(t)) |, \ \| \lambda \| < \epsilon,$$

the latter leading to $\sup_t | \ln(\lambda(t)/t) | < \epsilon$. Since

$$\ln(1 - 2\epsilon) < -\epsilon < \epsilon < \ln(1 + 2\epsilon), \ 0 < \epsilon < 1/4,$$

we have $\sup_t | \lambda(t) - t | < 2\epsilon$ and therefore $d(x,y) < 2d_1(x,y)$ whenever $d_1(x,y) < 1/4$. Thus each open d-ball contains an open d_1-ball.

Conversely, let $\epsilon \in (0, 1/4)$. By Lemma 6.4.5, there exists a partition $\{t_i\}$ of $[0,1]$ such that $t_{i+1} - t_i > \epsilon$ for all i and

$$sup\{| x(t) - x(s) | \mid s, t \in [t_{i-1}, t_i)\} < \alpha_x(\epsilon) + \epsilon, i \geq 1.$$

Let $d(x, y) < \epsilon^2$ for some $x, y \in D[0,1]$. Let $\mu \in \Lambda$ be such that

$$\begin{aligned} \sup_t \mid x(t) - y(\mu(t)) \mid &= \sup_t \mid x(\mu^{-1}(t)) - y(t) \mid < \epsilon^2, \\ \sup_t \mid \mu(t) - t \mid &< \epsilon^2. \end{aligned}$$

Let λ be such that it agrees with μ on $\{t_i\}$ and is piecewise linear. Since $\eta = \mu^{-1} \circ \lambda \in \Lambda$ (in particular, is increasing) and has $\{t_i\}$ for fixed points (i.e. $\eta(t_i) = t_i$ for all i), we have: $t \in [t_i, t_{i+1})$ if and only if $\eta(t) \in [t_i, t_{i+1})$. Thus

$$\begin{aligned} \mid x(t) - y(\lambda(t)) \mid &\leq \mid x(t) - x(\eta(t)) \mid + \mid x(\eta(t)) - y(\lambda(t)) \mid \\ &< \alpha_x(\epsilon) + \epsilon + \epsilon^2 \\ &< \alpha_x(\epsilon) + 4\epsilon. \end{aligned}$$

Since $\lambda(t_i) = \mu(t_i)$ for all i, $\sup_t \mid \mu(t) - t \mid < \epsilon^2$ and $t_{i+1} - t_i > \epsilon$ for all i, one has

$$\mid \lambda(t_i) - \lambda(t_{i-1}) - (t_i - t_{i-1}) \mid < 2\epsilon^2 < 2\epsilon \mid t_i - t_{i-1} \mid .$$

Since λ is piecewise linear, one then has

$$\mid \lambda(t) - \lambda(s) - (t - s) \mid < 2\epsilon \mid t - s \mid$$

for all $t, s \in [0,1]$, leading to

$$\ln(1 - 2\epsilon) \leq \ln[(\lambda(t) - \lambda(s))/(t - s)] \leq \ln(1 + 2\epsilon)$$

for $t \neq s$ in $[0,1]$. Since $\epsilon < 1/4$, one then has $\| \lambda \| < 4\epsilon$. Given $\delta > 0$, pick the ϵ above to satisfy in addition the condition $4\epsilon + \alpha_x(\epsilon) < \delta$. Then the foregoing implies that an open d_1-ball of radius δ contains a concentric open d-ball of radius ϵ^2. □

The Skorohod topology for $D[0,1]$ is simply the metric topology of d or, equivalently, d_1. The metric d is intuitively simpler of the two and leads to the following appealing characterization of convergence in $D[0,1]$, which captures the spirit of our earlier discussion: $x_n \to x$ in $D[0,1]$ if and only if there exist $\lambda_n \in \Lambda, n \geq 1$, such that $\lambda_n(t) \to t$ and $x_n(\lambda_n(t)) \to x(t)$ uniformly in t. The λ_n's thus represent the successive rescalings of the time axis. The problem with d is that it is not complete, as the following example shows: Let $x_n \in D[0,1], n \geq 1$, be defined by $x_n(t) = I\{\frac{1}{2} \leq t < \frac{1}{2} + \frac{1}{n}\}, t \in [0,1]$. Then it is easy to see that the infimum in the definition of $d(x_n, x_m), n, m \geq 1$, is attained at the $\lambda \in \Lambda$ which is linear on $[0, \frac{1}{2}]$, $[\frac{1}{2}, \frac{1}{2} + \frac{1}{m}]$ and $[\frac{1}{2} + \frac{1}{m}, 1]$ and maps these intervals respectively onto $[0, \frac{1}{2}], [\frac{1}{2}, \frac{1}{2} + \frac{1}{n}]$ and $[\frac{1}{2} + \frac{1}{n}, 1]$. This leads to $d(x_n, x_m) = \mid \frac{1}{n} - \frac{1}{m} \mid, n, m \geq 1$

(Exercise 6.18). Thus $\{x_n\}$ is Cauchy with respect to d. On the other hand, it is clear that if $y_n \to y$ in $D[0,1]$ and $y_n(t) \to a$ for some $t, a =$ either $y(t)$ or $y(t-)$. Using this in conjunction with the fact that $x_n \to I_{\{1/2\}}$ pointwise, one sees that $\{x_n\}$ does not converge in $D[0,1]$. This motivates the metric d_1 which is indeed complete.

Lemma 6.4.6 *The metric d_1 is complete.*

Proof Let $\{x_n\}$ be Cauchy with respect to d_1. Pick $\{k(m)\} \subset \{n\}$ such that $y_m = x_{k(m)}$ satisfy $d_1(y_m, y_{m+1}) < 2^{-m}, m \geq 1$. Thus there exist $\{\lambda_m\} \subset \Lambda$ such that $\| \lambda_m \| < 2^{-m}$ and

$$\sup_t \mid y_m(t) - y_{m+1}(\lambda_m(t)) \mid < 2^{-m}, m \geq 1.$$

From the proof of Lemma 6.4.4, recall that $\| \lambda \| < \epsilon$ implies $\sup_t \mid \lambda(t) - t \mid < 2\epsilon$ for $\epsilon \in (0, 1/4)$. Thus

$$\begin{aligned}\sup_t \mid \lambda_{n+m+1} \circ \lambda_{n+m} \quad \cdots \quad &\circ\lambda_{n+1} \circ \lambda_n(t) - \lambda_{n+m} \circ \ldots \circ \lambda_{n+1} \circ \lambda_n(t) \mid \\ &\leq \sup_t \mid \lambda_{n+m+1}(t) - t \mid \\ &\leq 2^{-(n+m)}, \quad n, m \geq 1.\end{aligned}$$

Thus for each $n \geq 1, \{\lambda_{n+m} \circ \ldots \circ \lambda_{n+1} \circ \lambda_n(.), m \geq 1\}$ are Cauchy in the supremum norm and therefore converge uniformly to some $\overline{\lambda}_n$. Since $\| \lambda_m \| < \infty, \{\lambda_m\}$ are continuous (Exercise 6.19), hence so is $\overline{\lambda}_n$ for $n \geq 1$. Clearly, $\{\overline{\lambda}_n\}$ are increasing with $\overline{\lambda}_n(0) = 0, \overline{\lambda}_n(1) = 1$ for $n \geq 1$. Thus $\{\overline{\lambda}_n\} \in \Lambda$. Since $\| f \circ g \| \leq \| f \| + \| g \|$ for $f, g \in \Lambda$,

$$\begin{aligned}\left| \ln \left[\frac{\lambda_{n+m} \circ \ldots \circ \lambda_{n+1} \circ \lambda_n(t) - \lambda_{n+m} \circ \ldots \circ \lambda_{n+1} \circ \lambda_n(s)}{t - s} \right] \right| \\ \leq \| \lambda_{n+m} \circ \ldots \circ \lambda_{n+1} \circ \lambda_n \| \\ \leq \| \lambda_{n+m} \| + \ldots + \| \lambda_n \| \\ \leq 2^{-(n-1)}.\end{aligned}$$

Letting $m \to \infty$ on the left hand side, we have $\| \overline{\lambda}_n \| \leq 2^{-(n-1)}$. Now $\overline{\lambda}_n = \overline{\lambda}_{n+1} \circ \lambda_n$ and thus

$$\begin{aligned}&\sup_t \mid y_n(\overline{\lambda}_n^{-1}(t)) - y_{n+1}(\overline{\lambda}_{n+1}^{-1}(t)) \mid \\ &= \sup_t \mid y_n(t) - y_{n+1}(\lambda_n(t)) \mid < 2^{-n}, n \geq 1.\end{aligned}$$

Thus $\{y_n(\overline{\lambda}_n^{-1}(.))\}$ is Cauchy in supremum norm and therefore converges uniformly to a function $x(.) : [0,1] \to \boldsymbol{R}$. Since uniform limits of *cadlag* functions are *cadlag* (Exercise 6.20), $x \in D[0,1]$. Since $y_n(\overline{\lambda}_n^{-1}(.)) \to x(.)$ uniformly and $\| \overline{\lambda}_n^{-1} \| = \| \overline{\lambda}_n \| \to 0, d_1(y_n, x) \to 0$. The claim follows. □

Theorem 6.4.2 *$D[0,1]$ is a Polish space.*

Proof We only need to establish the separability of $D[0,1]$. Let $S \subset D[0,1]$ be the countable set of functions that are constant on intervals $\left[\frac{i}{k}, \frac{i+1}{k}\right), 0 \leq i < k$, for some $k \geq 1$ and take only rational values. Let $x \in D[0,1], \epsilon > 0$. As in the proof of Lemma 6.4.5, pick $0 = t_0 < t_1 < \ldots < t_n = 1$ such that $\sup\{\mid x(s) - x(t) \mid: \ s, t \in [t_i, t_{i+1})\} < \epsilon$ for $0 \leq i < n$. Let $y \in D[0,1]$ be constant on each interval $[t_i, t_{i+1})$ with $y(t_i) = x(t_i)$ for each i. Then $d(x,y) < \epsilon$. Pick rationals $r_i, 0 \leq i \leq n$ such that $\max_i \mid x(t_i) - r_i \mid < \epsilon$ and let $z \in D[0,1]$ be constant on each $[t_i, t_{i+1})$ with $z(t_i) = r_i$ for each i. Then $d(y,z) < \epsilon$. Finally, let $k \geq 1$ be large enough so that $k^{-1} < min(\epsilon, min(t_{i+1} - t_i)/4)$. Let $A = \left\{\frac{j}{k}, 0 \leq j \leq k\right\}$ and $s_i = min\{s \in A \mid s \geq t_i\}, 0 \leq i \leq n$. Let $\lambda \in \Lambda$ be the piecewise linear function that maps s_i to t_i for $0 \leq i \leq n$ and is linear on each interval $[s_i, s_{i+1}]$. Let $u \in D[0,1]$ be constant on each interval $[s_i, s_{i+1}]$ with $u(s_i) = r_i$ for each i. Then $u(.) = z(\lambda(.))$, leading to $d(z,u) < \epsilon$. Together, these imply that $d(x,u) < 3\epsilon$. Since $\epsilon > 0$ was arbitrary and $u \in S$, it follows that S is dense in $D[0,1]$. □

As in the continuous case, one would like to view a *cadlag* process as a $D[0,1]$-valued random variable. The next lemma paves the way for this. Let $G \subset [0,1]$ be countable dense and contain 1.

Lemma 6.4.7 *The σ-field $\mathcal{G}$ of subsets of $D[0,1]$ generated by sets of the type $\{x(.) \mid x(t_i) \in A_i, 1 \leq i \leq n\}$ for some $n \geq 1, \{t_i\} \subset G$ and $\{A_i\}$ Borel in $\boldsymbol{R}$, coincides with $\mathcal{B} =$ the Borel σ-field of $D[0,1]$.*

Proof Let $t \in [0,1]$. If $x_n \to x$ in $D[0,1]$ and $t = 1, x_n(t) \to x(t)$ and thus the map $x \to x(t)$ is continuous, therefore measurable. If $t < 1$, for any $\epsilon > 0$, the dominated convergence theorem leads to

$$h_\epsilon(x_n) = \frac{1}{\epsilon}\int_t^{t+\epsilon} x_n(s)ds \to h_\epsilon(x) = \frac{1}{\epsilon}\int_t^{t+\epsilon} x(s)ds.$$

Thus $x \to h_\epsilon(x)$ is continuous. Since $x(t) = \lim_{\epsilon \to 0} h_\epsilon(x)$, the map $x \to x(t)$ is measurable. Thus $\mathcal{G} \subset \mathcal{B}$.

Conversely, let $x \in D[0,1], r > 0, B = \{y \in D[0,1] \mid d_1(x,y) < r\}$. We shall prove that $B \in \mathcal{G}$, thereby proving that $\mathcal{B} \subset \mathcal{G}$ and hence $\mathcal{B} = \mathcal{G}$. Let $G = \{t_1, t_2, \ldots\}$ with $t_1 = 1$. For $0 < \epsilon < r, k \geq 1$, define

$$\begin{aligned}
A_k(\epsilon) &= \{y \in D[0,1] \mid \exists \lambda \in \Lambda \ \ with \parallel \lambda \parallel < r - \epsilon \text{ and} \\
&\quad \max_{1 \leq i \leq k} \mid y(t_i) - x(\lambda(t_i)) \mid < r - \epsilon\}, \\
H_1 &= \{(x(\lambda(t_1)), \ldots, x(\lambda(t_k))) \mid \lambda \in \Lambda, \parallel \lambda \parallel < r - \epsilon\} \subset \boldsymbol{R}^k, \\
H_2 &= \{(a_1, \ldots, a_k) \in \boldsymbol{R}^k \mid \exists (b_1, \ldots, b_k) \in H_1 \text{ such that} \\
&\quad \max_i \mid a_i - b_i \mid < r - \epsilon\}.
\end{aligned}$$

Let $\pi : D[0,1] \to \boldsymbol{R}^k$ denote the $\mathcal{G}$-measurable map $y(\cdot) \to (y(t_1), \ldots, y(t_k))$. H_2 is open and $A_k(\epsilon) = \pi^{-1}(H_2)$, hence $\mathcal{G}$-measurable. Clearly, $B \subset B' = \bigcup_\epsilon \bigcap_{k=1}^\infty A_k(\epsilon)$, the union being over all rational ϵ in $(0, r)$. On the other hand, for $y \in \bigcap_{k=1}^\infty A_k(\epsilon)$, pick $\lambda_k \in \Lambda, k \geq 1$, such that $\| \lambda_k \| < r - \epsilon, \max_{1 \leq i \leq k} | y(t_i) - x(\lambda_k(t_i)) | < r - \epsilon$. Let $\{X_k\}$ be $[0,1]$-valued random variables such that $P(X_k \leq a) = \lambda_k(a)$ for $a \in [0,1], k \geq 1$. Since $\boldsymbol{P}([0,1])$ is compact, $X_{k(n)} \to X_\infty$ in law for some $[0,1]$-valued random variable X_∞ and some subsequence $\{k(n)\}$ of $\{k\}$. Letting $\lambda(a) = P(X_\infty \leq a)$ for $a \in [0,1]$, it then follows that $\lambda_{k(n)}(t) \to \lambda(t)$ for all continuity points of λ. Also, if s, t are distinct continuity points of λ,

$$\begin{aligned} | \ln[(\lambda(t) - \lambda(s))/(t - s)] | &= \lim_{n \to \infty} | \ln[(\lambda_{k(n)}(t) - \lambda_{k(n)}(s))/(t - s)] | \\ &\leq r - \epsilon. \end{aligned}$$

Since λ can have only jump discontinuities at at most countably many points, it follows from the above inequality that λ must be continuous at all points. Furthermore, it also implies that λ is strictly increasing. Thus $\lambda \in \Lambda$. The above inequality also leads to $\| \lambda \| \leq r - \epsilon$. Since $t_1 = 1$, it follows from the definition of $A_k(\epsilon)$ that $| y(t_1) - x(t_1) | < r - \epsilon$. For $i > 1$, our choice of $\{\lambda_k\}$ implies that $| y(t_i) - x(\lambda_k(t_i)) | < r - \epsilon$ for $k \geq i$. Letting $k \to \infty$ along $\{k(n)\}$, we have

$$| y(t) - x(\lambda(t)) | \wedge | y(t) - x(\lambda(t)-) | \leq r - \epsilon, \ t \in G.$$

Since G is dense in $[0,1], \sup_t | y(t) - x(\lambda(t)) | \leq r - \epsilon$. Together with $\| \lambda \| \leq r - \epsilon$, this implies that $y \in B$. Thus $B' \subset B$ and therefore $B = B'$, implying $B \in \mathcal{G}$. This completes the proof. □

Let $X_t, t \in [0,1]$, be a real-valued process satisfying the conditions (i) and (ii) (or (ii′)) of Lemma 6.4.2. Let $\{\overline{X}_t\}$ be its *cadlag* version and for each sample point ω, let $\overline{X}(\omega)$ denote the entire trajectory $t \to \overline{X}_t(\omega)$ viewed as an element of $D[0,1]$.

Theorem 6.4.3 *$\overline{X}$ is a $D[0,1]$-valued random variable.*

Proof Without any loss of generality, we may suppose that the countable dense set $D \subset [0,1]$ of Lemma 6.4.2 contains 1. Set $G = D$ in the preceding lemma. The claim now follows from the above lemma and the fact that $\overline{X}_t = X_t$ for $t \in D$, the latter statement being implicit in the proof of Lemma 6.4.2 (cf. Lemma 6.3.1). □

In conjunction with Theorem 6.4.2, this allows us to use the machinery of Chapter 2 for studying *cadlag* processes.

6.5 Examples of Stochastic Processes

An in-depth study of continuous-time stochastic processes is beyond the scope of this book. By way of a glimpse of the subject and also as an epilogue for this book, here's a list of some important classes of stochastic processes:

(1) **Independent increment processes** — These are processes $\{X_t\}$ that satisfy: for $t_1 < t_2 < t_3 < \ldots$, the random variables $X_{t_2} - X_{t_1}, X_{t_3} - X_{t_2}, \cdots$, are independent. These processes may be viewed as the continuum analogues of sums of independent random variables and often serve as the basic building blocks for more general stochastic processes.

(2) **Martingales** — $\{X_t\}$ is a martingale with respect to an increasing family of σ-fields $\{\mathcal{F}_t\}$ if X_t is integrable and $\mathcal{F}_t$-measurable for all t and for $t \geq s, E[X_t/\mathcal{F}_s] = X_s$ a.s. These processes not only inherit much of the theory we developed for discrete time martingales in Chapter 3, but have a lot more richer theory of their own, a high point of which is the "stochastic calculus". Further refinements and extensions of these include continuous time super/submartingales, local martingales, semimartingales, etc.

(3) **Markov processes** — $\{X_t\}$ is a Markov process if $X(t + \cdot)$ is conditionally independent of $\{X_s, s \leq t\}$ given X_t for each t. As in the case of discrete time Markov chains, these serve as stochastic analogs of "dynamical systems." Important refinements include strong Markov processes, diffusion processes and so on.

(4) **Jump processes** — These are piecewise constant *cadlag* processes. An important subclass is the point processes which are integer-valued and have nondecreasing trajectories. These are important in queueing theory, branching processes, etc. Further refinements of jump processes include piecewise deterministic processes, random measures, etc.

(5) **Stationary processes** — $\{X_t\}$ is a stationary process if for any finite collection $t_1 < t_2 < \ldots < t_n$ of time instants, the distributions of $[X_{t_1}, ..., X_{t_n}]$ and $[X_{t_1+h}, ..., X_{t_n+h}]$ agree for any $h > 0$. Further refinements and extensions include ergodic processes, wide sense stationary processes, etc.

(6) **Gaussian processes** — $\{X_t\}$ is a Gaussian process if its finite dimensional marginals are Gaussian. Because of the analytic advantages of working with Gaussian distributions and reasonableness of the Gaussian approximation in many real world situations, this rather special class of processes has been extensively studied.

Some examples of continuous time stochastic processes are:

(a) **Brownian motion** — The Brownian motion $\{B_t\}$ is an independent increment process satisfying: For any $t > s, W(t) - W(s)$ is Gaussian with mean zero and variance $t - s$. An application of Theorem 6.3.1 shows that it has a continuous version (Exercise 6.21). This sample path continuity is usually built into the definition of a Brownian motion. Obviously a Gaussian process, it is also a martingale with respect to $\sigma(B_s, s \leq t), t \geq 0$, and a Markov process as well (Exercise 6.22).

(b) **Poisson process** — The Poisson process $\{N_t\}$ is an independent increment process satisfying: For $t > s, N_t - N_s$ has a Poisson distribution with parameter $\lambda(t-s)(i.e., P(N_t - N_s = n) = (t-s)^n \lambda^n exp(-\lambda(t-s))/n!$ for $n \geq 0$). An application of Theorem 6.4.1 shows that it has a *cadlag* version (Exercise 6.23)). The *cadlag* property of sample paths is usually built into the definition of a Poisson process. Obviously a point process, it is also a Markov process and $N_t - \lambda t, t \geq 0$, is a martingale with respect to $\sigma(N_s, s \leq t), t \geq 0$ (Exercise 6.24).

(c) **Ornstein-Uhlenbeck process** — The integral equation

$$X_t = X_0 - \frac{1}{2}\int_0^t X_s ds + B_t, \; t \geq 0,$$

where $\{B_t\}$ is a Brownian motion, can be explicitly solved using the "variation of constants" formula. (This explicit solution involves the so called stochastic integral, which is beyond the scope of this book.) The a.s. unique solution $\{X_t\}$ thereof defines a continuous sample path process known as the Ornstein-Unlenbeck process. This process is quite important in physics. It turns out to be a Markov process (in fact, a "diffusion") and is Gaussian if X_0 is. In particular, if X_0 is Gaussian with mean zero and variance 1, it becomes a stationary (in fact, an "ergodic") process.

Each of the aforementioned classes of continuous time stochastic processes is a thriving subdiscipline of the vast and everexpanding theory of stochastic processes. A reader who has studied this book carefully should be quite well-equipped to venture into one or more of these areas. For the reader so inclined, my personal advice would be to read the two volume classic [40,48].

6.6 Additional Exercises

(6.25) Show that in Lemma 6.3.1, convergence in probability may be replaced by convergence in law. (Hint: Show that for real random variables defined on a common probability space, convergence in law to a constant implies convergence in probability to the same.)

(6.26) Give an example of a *cadlag* process $\{X_t\}$ which is a.s. continuous at every point (i.e., $X_s \to X_t$ a.s. whenever $s \to t$), but has discontinuous sample paths a.s.

(6.27) Let $\{X_t\}$ be a Gaussian process satisfying $E[(X_t - X_s)^2] \leq C \mid t-s \mid^b$ for some $C, b > 0$. Show that it has a continuous version.

(6.28) Let $X_t, t \in D = \{(t_1, ..., t_n) \in \boldsymbol{R}^n \mid 0 \leq t_i \leq 1\}$ be a family of real random variables satisfying

$$E[\mid X_t - X_s \mid^a] \leq C \mid t - s \mid^{n+b}, t, s \in D,$$

for some $a, b, C > 0$. Show that there exists another family $\{\overline{X}_t, t \in D\}$ of real random variables such that $X_t = \overline{X}_t$ a.s. and $t \in D \to \overline{X}_t$ is continuous a.s.

(6.29) Show that the conclusions of Theorem 6.3.1 may fail for $c = 0$. (Hint: Consider the Poisson process.)

(6.30) Show that the condition (6.1) of Theorem 6.3.1 implies the condition (6.2) of Theorem 6.4.1 for suitably redefined a, b.

(6.31) Give an example of a *cadlag* function on $[0, 1]$ whose points of discontinuity are dense in $[0, 1]$.

(6.32) For Λ as in Section 6.4 above, show that for Lipschitz continuous (and hence a.e. differentiable, by Rademacher's theorem) $\lambda \in \Lambda, \| \lambda \| = esssup_{t \in [0,1]} \mid \ln \lambda'(t) \mid$ where $\lambda'(\cdot)$ is the derivative of λ, defined a.e.

References

[1] V. Anantharam and P. Tsoucas. 1989. A proof of the Markov chain tree theorem. *Statistics and Probability Letters*, 8:189–192.

[2] H. Bergstrom. 1982. *Weak Convergence of Measures.* Academic Press, New York.

[3] P. Billingsley. 1968. *Convergence of Probability Measures.* John Wiley, New York.

[4] N.H. Bingham. 1986. Variants on the law of the iterated logarithm. *Bull. London Math. Soc.*, 18:433–467.

[5] D. Blackwell and L. Dubins. 1983. An extension of Skorohod's almost sure representation theorem. *Proceedings of American Mathematical Society*, 89:691–692.

[6] D. Blackwell and L. Dubins. 1962. Merging of opinions with increasing information. *Annals of Mathematical Statistics*, 33:882–886.

[7] L. Breiman. 1968. *Probability.* Addison-Wesley, Reading, MA.

[8] Y.S. Chow and H. Teicher. 1978. *Probability Theory — Independence, Interchangeability and Martingales.* Springer-Verlag, New York.

[9] K.L. Chung. 1974. *A First Course in Probability Theory.* Academic Press, New York.

[10] F.N. David. 1962. *Games, Gods and Gambling.* Charles Griffin, London.

[11] C. Dellacherie and P.-A. Meyer. 1978. *Probabilities and Potential.* North Holland, Amsterdam.

[12] C. Dellacherie and P.-A. Meyer. 1982. *Probabilities and Potential B — Theory of Martingales.* North Holland, Amsterdam.

[13] J.D. Deuschel and D.W. Stroock. 1989. *Large Deviations.* Academic Press, New York.

[14] J.L. Doob. 1989. Kolmogorov's early work on convergence theory and foundations. *Annals of Probability*, 17:815–821.

[15] J.L. Doob. 1953. *Stochastic Processes.* Wiley, New York.

[16] L. Dubins. 1983. Bernstein-like polynomial approximation in higher dimensions. *Pacific Journal of Mathematics*, 109:305–311.

[17] R.S. Ellis. 1985. *Entropy, Large Deviations and Statistical Mechanics.* Springer-Verlag, New York.

[18] N. Etemadi. 1981. An elementary proof of the strong law of large numbers. *Z. Wahrsch. verw. Gebiete*, 55:119–122.

[19] S. Ethier and T.G. Kurtz. 1986. *Markov Processes — Characterization and Convergence.* Wiley, New York.

[20] T.L. Fine. 1973. *Theories of Probability.* Academic Press, New York.

[21] D. Freedman. 1983. *Markov Chains.* Springer-Verlag, New York.

[22] I.I. Gikhman and A.V. Skorohod. 1969. *Introduction to the Theory of Stochastic Processes.* W.B. Saunders, Philadelphia.

[23] B.V. Gnedenko and A.N. Kolmogorov. 1954. *Limit Distributions for Sums of Independent Random Variables.* Addison-Wesley, Reading, MA.

[24] I. Hacking. 1975. *The Emergence of Probability.* Cambridge University Press, London.

[25] D.L. Isaacson and R.W. Madsen. 1976. *Markov Chains: Theory and Applications.* John Wiley, New York.

[26] E. Jaynes. 1968. Prior probability. *IEEE Transactions on Systems Science and Cybernetics*, 4:227–241.

[27] J. Kemeny, J.L. Snell, and A.W. Knapp. 1976. *Denumerable Markov Chains.* Springer-Verlag, New York.

[28] M.J. Klass. 1975. A survey of the $\frac{S_n}{n}$ problem. In M.L. Puri, editor, *Statistical Inference and Related Topics, Vol. 2: Proc. of the Summer Research Institute on Statistical Inference for Stochastic Processes,* Academic Press, New York.

[29] H.H. Kuo. 1975. *Gaussian Measures on Banach Spaces. Lecture Notes in Maths., No. 463,* Springer-Verlag, New York.

[30] R.B. Lindsay and H. Margeneau. 1957. *Foundations of Physics.* Dover, New York.

[31] M. Loeve. 1977. *Probability Theory I.* Springer-Verlag, New York, second edition.

[32] E. Lukacs. 1960. *Characteristic Functions.* Charles Griffin, London.

[33] L.E. Maistrov. 1974. *History of Probability.* Academic Press, New York.

[34] J. Neveu. 1965. *Mathematical Foundations of the Calculus of Probability.* Holden-Day, San Fransisco.

[35] E. Nummelin. 1984. *General Irreducible Markov Chains and Nonnegative Operators.* Cambridge University Press, Cambridge.

[36] S. Orey. 1971. *Limit Theorems for Markov Chain Transition Probabilities.* Van Nostrand, London.

[37] K.R. Parthasarathy. 1977. *Introduction to Probability and Measure.* McMillan (India), New Delhi.

[38] K.R. Parthasarathy. 1967. *Probability Measures on Metric Spaces.* Academic Press, New York.

[39] D. Pollard. 1984. *Convergence of Stochastic Processes.* Springer-Verlag, New York.

[40] L.C.G. Rogers and D. Williams. 1987. *Diffusions, Markov Processes and Martingales II: Ito Calculus.* Wiley, Chichester.

[41] L.J. Savage. 1954. *Foundations of Statistics.* Wiley, New York.

[42] L. Schwartz. 1976. *Disintegration of Measures.* Tata Institute of Fundamental Research, Bombay.

[43] G. Shafer. 1990. Unity and diversity of probability. *Statistical Science*, 5:435–444.

[44] V. Strassen. 1966. A converse to the law of the iterated logarithm. *Z. Wahrsch. verw. Gebiete*, 4:265–268.

[45] R.L. Tweedie. 1974. A representation for invariant measures of transient Markov chains. *Zeit. Wahr.*, 28:99–112.

[46] S.R.S. Varadhan. 1984. *Large Deviations and Applications.* Society for Industrial and Applied Mathematics, Philadelphia.

[47] M.J. Wichura. 1970. On the construction of almost uniformly convergent random variables with given weakly convergent image laws. *Annals of Mathematical Statistics*, 41:284–291.

[48] D. Williams. 1979. *Diffusions, Markov Processes and Martingales I: Foundations.* Wiley, Chichester.

[49] E. Wong and B. Hajek. 1985. *Stochastic Processes in Engineering Systems.* Springer-Verlag, New York.

[50] H. Zhenting and G. Qingfeng. 1988. *Homogeneous Denumerable Markov Processes.* Springer-Verlag, Berlin - Heidelberg.

Index